CLIMATE CHANGE
SIMPLIFIED

A COMPREHENSIVE GUIDE TO GLOBAL WARMING AND SUSTAINABLE LIVING WITH 101 ESSENTIAL TIPS FOR A GREENER FUTURE AND REDUCED CARBON FOOTPRINT

ALEXA INGRAM

Dedicated to my daughter Sasha, the future custodian of our planet.

Embrace the past without regret, for it cannot be altered.

Transform ripples into waves, and remember,

your potential knows no bounds.

TABLE OF CONTENTS

INTRODUCTION

My passion for environmental issues first started while planning a trip to the Greek islands to celebrate my beloved daughter's 16th birthday. While I never really understood the tradition of having a Sweet 16, I wanted to create a memorable experience for her rather than shelling out money on a big bash. Little did I know at the time, but this expedition would also represent a turning point in my life. It would be the first time that I really thought deeply about the hidden environmental costs of our day-to-day lives.

As I sat in front of my laptop, booking ferry tickets from the port of Piraeus on the Greek mainland to the islands, a strange thing caught my eye—a rating system that appeared next to each ticket price. To my surprise, there it was, stated right before my eyes, a number that spoke volumes: the carbon impact of each ferry ticket. Suddenly, the tickets represented more than just a seat booking for the fab vacation we were about to embark on; they became messengers of truth, unmasking the hidden ecological impacts of our travel choices.

At first, I thought this carbon rating system was just a small element of the website's design, perhaps a fleeting gesture on the part of the ferry operator toward trying to appear more "green." After all, these days, lots of businesses have started these types of initiatives to try and appear more climate-friendly. As I continued scrolling through the information on the website, I found out more.

The ferry options varied not just in price and convenience but also in ecological footprint. The faster, more luxurious ferries had a higher emission tag, while the slow boats that putter along the Aegean at a more leisurely pace had a lesser environmental impact. It became clear to me that the choices we make as travelers, vacationers, and world explorers ultimately contribute to the fate of our planet.

While the carbon footprint of a single ferry ride may seem trivial, think about people who travel by commercial air frequently, or worse yet—private jets! Our roles as individuals, as societies, and as a collective human race, and the ways we choose to travel for business or pleasure are intertwined.

As far as my personal choices, I commute from New Jersey to New York City, and while there are a number of options for crossing the Hudson, I almost always choose the ferry. Although the PATH train is also nearby, I choose the waterway almost all the time because I'd rather spend my morning commute on a gentle cruise than underground– yes, even on those blisteringly cold, windy winter days! I love to get fresh air; what else can I say? The "aha moment" for me, when I was struggling to figure out which commuting option would work best for my personal needs and preferences, came from realizing this: I spent so much on an electric car, and I don't even have any idea what the footprint is from making this choice! This really got me thinking. Since then, I've dedicated a ton of time to trying to be aware of the ecological footprint of my daily choices. We have nutrition facts on the back of all types of food items at the supermarket: Why not have a similar, easy-to-understand guide system that lets us know how much energy is expended by our transportation choices?

I've realized that as a citizen of a highly developed country with access to luxuries and conveniences, I'm in no way absolved of my own role in contributing to the widening carbon toll of our society. Those of us with more privilege bear an even greater responsibility to be the catalysts of change and must take a crucial role in spearheading efforts to protect the planet for future generations.

The Greece trip with my daughter gave me time to reflect and taught me ways of framing my own privilege while thinking about ways to confront it head-on. Upon recognizing the limited information available to the average consumer, I came to the realization that merely driving an electric vehicle and relying on solar energy to power my home fell short and this realization motivated me to make substantial changes.

Now, since you've gone ahead and purchased this book, I can see that you also have that yearning within you: the desire for real solutions, for actionable steps that can bring about positive change. I assure you that this book is not just another catchy title that plays off buzzwords and trends. It's a journey towards ecological enlightenment, empowering you to make meaningful and sustainable choices in your own daily life!

By reading this book, you'll gain powerful insights into the true impact our decisions have on the planet. You'll be equipped with knowledge about eco-friendly alternatives and sustainable practices that can be easily integrated into your daily routine. The book can serve as a guide for you, offering shortcuts to understanding complex environmental concepts and outlining concrete ways you can start taking impactful steps without feeling overwhelmed.

Throughout the pages of this book, my greatest hope is that it can help you find the strength to align your values with your actions while also making it easier to process news stories about climate catastrophes without feeling overwhelmed by anxiety. Hopefully, it'll also help you feel more connected to a greater community of like-minded individuals, all striving for a greener, more sustainable future. Only together can we assume the collective responsibility of safeguarding the planet for generations to come.

So, my fellow ecological enthusiast, let's set off along this path together. It's time to embrace the power of knowledge and conscious decision-making!

Action has been called for, and through our shared commitment, we, together, can be the catalyzing force for real change. Global warming has turned into a crisis of stunning scale and reach and can no longer be ignored. Raising awareness and making small lifestyle adjustments in our own households can hopefully start the much-needed ripples that we need to send out to spread the word and inspire others to take action along with us.

THE ECO RIPPLE SYSTEM

There's a transformative way of being and living that I like to call the Eco Ripple System. Based on the idea of the ripple effect, this system is one that can inspire us to be more conscious in our personal decision-making processes and, hopefully, stoke a collective dispersion of positive change throughout our own communities that ripple out to the greater world. Imagine a world where every individual becomes a harbinger of positive change, their actions mirroring a steadfast commitment to protecting our planet.

Over the last few years, there's been a distinct uptick in interest in the environment and being "green." Ordinary people who previously haven't thought much about their climate impact and not-so-climate-concerned businesses have suddenly transformed into eco-warriors, reshaping their lives to align with the prevalent sustainable practices of our times. While not all these gestures are sincere (and we'll be talking about that later), they do represent a big, positive step in the right direction. From bustling metropolises to remote villages, the ripple effect of conscious choices is something that can be felt collectively at this moment, giving us hope that we can restore this precious planet.

Picture yourself standing at the epicenter of this global move-ment for change, your daily decisions infused with a newfound sense of purpose, knowing that every seemingly small action actually has the power to make a significant impact when coupled with the combined efforts of others. This is what the Eco Ripple System is all about. With each step, you, too, can contribute to the sweeping winds of change that are beginning to echo across continents, transforming the trajectory of climate change and fighting for the preservation of the world for generations to come.

With the Eco Ripple System comes paradigm shifts. What it encapsulates is a holistic approach to sustainability, presenting a comprehensive guide that empowers you to enact personally meaningful change in your life. No longer will you feel over-whelmed by the scale of the challenge; instead, you'll be armed with empowering tools that'll help you embrace your role as an eco-champion with confidence and clarity.

I promise that the journey we undertake throughout these pages won't be arduous or monotonous. Despite the heavy subject matter, it'll also be filled with enlightenment, inspira-tion, and the camaraderie of a community united in a shared purpose. As you read on, you may even feel an innate inner resonance or hear a subtle whisper that tickles the hair on the back of your neck, affirming, "This is the right book for me." The Eco Ripple System is not just another book on environ-mentalism; rather, it's a gateway to a brighter future where each of us, together, holds the power to reshape the world for the better.

1

CLIMATE CONSCIOUSNESS

I magine a world where the Earth's average temperature has risen substantially since the beginning of industrialization, where the landscapes we once knew have begun to transform and sometimes break under the relentless stress of warming trends. That reality is happening now! While 1.1 degrees Celsius might not seem like a whole lot to you, this climbing global temperature has already begun to rapidly reshape our natural world.

While some corners of this fragile planet may see sporadic cooling effects at some (unknowable) future time, the inexorable march towards hotter days is an undeniable fact: One that's rewriting the prevailing narrative of global progress and carving out the eventual fate of human civilization. This warming trend, however patchy and uneven, is currently sweeping the world, bringing a number of destructive and disturbing trends with it.

Our beautiful, blue and green planet, once marked by a delicate balance of species and ecosystems, is now faced with a heavily adulterated, artificial, and ever-more precarious new equilibrium brought on by the heavy impact of human civilization. The shift in global temperature we're witnessing is emitting shockwaves, ushering in consequences that cascade through regions, wreaking havoc on coastlines, natural geographic features and inland waterways, natural wildlife habitats, and threatening to uproot the lives of human populations along the way. A cacophonous symphony of change is playing out before us, and as the warning trumpets blast, its crescendo resonates far beyond the boundaries of the natural world.

This chapter will take us into the heart of climate consciousness, where we'll be diving deep into the mechanics of climate change, grasping the technical factors surrounding greenhouse gas emissions, and seeking to understand and address the ever-more-urgent need to drastically shrink the carbon impact of our modern-day civilizations.

As we go through the realities being confronted today, we'll uncover hidden sources of emissions, trace the origins of global warming, and confront the tangible impacts of the massive global shift we're currently undergoing and its many impacts that'll come to define the future.

From flickers of awareness to a blaze of understanding, the Eco Ripple System begs us to listen and confront these realities. Together, we'll explore how our actions, or rather our collective inactions, are becoming indelible marks on the surface of what astronomer and science writer Carl Sagan (1994) endearingly referred to as the "pale blue dot": Earth.

Buckle up because, in the pages that follow, we'll lay bare, sometimes in stark terms, the driving forces behind climate change. Keep in mind, however: Being blunt about the realities we face doesn't mean we can't embrace the imperative to alter our trajectory. So, onwards we go on our path toward comprehension and compassion, driven by the building collective sense of urgency to safeguard the world we share.

UNDERSTANDING CLIMATE CHANGE

In order to understand the full picture of climate change and its ramifications, we first have to start by unraveling its inherent complexities, from the mechanics of the greenhouse effect to the far-reaching implications of global warming. It's important to grasp the interplay of the various factors that are currently reshaping our world and instill in ourselves a deep sense of urgency to mitigate their effects.

What Is the Greenhouse Effect?

Picture Earth as a colossal terrarium. Just as the glass encases our little imaginary garden, acting like a protective shell, trapping heat from the sun and holding it within, so does the Earth's atmosphere. The atmosphere, composed of a delicate balance of gasses, is one of the essential characteristics of Earth that allows it to support life and maintain a certain equilibrium. In fact, the phenomenon known as the greenhouse effect, where certain gasses trap heat in the atmosphere, is the same factor that keeps our planet warm enough to support a diverse range of animal life and ecosystems. Without the atmosphere and the greenhouse gasses trapped within it, life would cease to exist.

"But greenhouse gasses are bad, right?" you might ask; well, yes, in fact, they are. Here we'll explain:

Human activity on our planet, the things we do on a day-to-day basis, and the systems we participate in and support are the primary cause of an increasing greenhouse effect. An increasing greenhouse effect is bad news for us and all life on Earth.

It happens like this: When the fragile equilibrium of the natural greenhouse effect is thrown off-balance through the impact of human activity, it can easily tip the scales toward an ever-warming climate. This change can disrupt the life-supporting characteristics of our planet. An increasing greenhouse effect due to the human release of greenhouse gasses has been recognized by scientists at least since the 1980s, but its origins can be traced back to the early industrial age.

As more and more carbon dioxide (CO_2) and other greenhouse gasses are released into the air, mainly through the burning of fossil fuels, the greenhouse effect causes higher levels of heat to get trapped inside the atmosphere, creating a compounded global warming effect.

Okay, so now that that part's clear, you might be wondering: What exactly are the different types of greenhouse gasses, and how many of them are there?

Carbon dioxide (CO2), methane (CH4), nitrous oxide (N2O), and a number of different fluorinated gasses comprise what we consider to be greenhouse gasses. While these gasses are essential atmospheric actors that enable our planet to retain warmth, they're also the same gasses that have the potential to upset the fragile balance when released into the atmosphere in mass quantities. Where do these mass quantities of excess gasses come from? Some of the answers may surprise you.

Where Emissions Come From Today

While the towering smokestacks of fossil-fuel-burning plants and the grimy chimneys of heavy fossil-fuel-reliant industries paint an obvious picture of environmental impact, there are also many lesser-known culprits. It's also worth taking a closer look at some of these unexpected sources that also contribute significantly to our global carbon footprint.

Fast Fashion

This industry, which thrives globally, is based on the idea of rapidly copying and bringing to market trendy designer clothing styles, then offering them at bargain basement prices. This cycle puts pressure on the consumer to continuously update their wardrobes. It goes without saying, but this practice, in which clothing is treated as practically disposable, ends up creating huge amounts of waste. Additionally, the trend-based cycles that drive the rapid production of new goods often rely on low-cost materials and labor, two other factors that have their own respective tolls on the environment and human rights.

From the manufacturing processes that rely on energy-intensive practices to the transportation and logistics networks that bring these garments to racks of stores across the globe, fast fashion's carbon footprint is broad and expansive, representing 10% of all pollution across the globe (Clarke, 2021). Most shoppers probably wouldn't even consider it, but their habits of buying cheap, mass-produced clothing have an even greater impact on the environment than the air travel industry (Stallard, 2022).

Glass Containers

While glass containers may seem like a more eco-friendly choice than plastic bottles, believe it or not, the same amount of energy it takes to manufacture five plastic containers goes into just one glass bottle (Antos, 2020). Undoubtedly, plastic pollution is a major environmental issue, and there are growing concerns about microplastic contamination everywhere, including our oceans. However, many plastic goods do, in fact, end up getting recycled. While glass is also often recycled, it's the heat factor that we have to look at when assessing the carbon impact.

Glass production–and remelting of recycled glass–involves heating raw materials at higher temperatures, which demands substantial energy inputs. This energy-intensive process, often fueled by fossil fuels, releases greenhouse gasses into the atmosphere. Additionally, the transportation of heavy glass containers from manufacturing facilities to stores and homes adds to their overall carbon footprint.

Agriculture, Rice Farming in Particular

One of the most important global staple foods is rice, but did you know that it has a hidden ecological impact? The cultivation of rice emits the greenhouse gas methane. Rice cultivation is thought to be responsible for up to 10% of methane emissions worldwide (Umali-Deininger, 2022). Talk about a big footprint!

What happens is this: Rice paddies need to be flooded to provide the necessary environment for rice to grow. This flooding makes these areas essentially anaerobic environments, meaning that there's a lack of oxygen in the ground. While these flooded anaerobic fields aren't conducive to most forms of life, certain bacteria can thrive under these conditions. The bacteria that grow in these waterlogged soils produce methane, and as methane escapes into the atmosphere, it contributes to the greenhouse effect.

While rice is a delicious and healthy grain, other staple carbohydrate food groups, like wheat and potatoes, have significantly lower carbon footprints. While all grains, including wheat, use substantial amounts of water resources, potatoes and other tubers boast the lowest overall impact of any of these common starch food groups.

Streaming Platforms and Server Farms

Streaming movies, music, and videos necessitates the operation of vast server farms to store and deliver all types of digital content, which consumes substantial amounts of energy. This energy demand stems from the need to power servers, cool data centers, and transmit cloud data across networks.

The production of the energy required to run streaming platforms and server farms often relies on fossil fuels, contributing to greenhouse gas emissions. Believe it or not, the electricity that fueled YouTube alone emitted a staggering "10 million metric tons of CO2-equivalent gasses globally in 2016" (CNN, 2020)—a figure equivalent to the annual emissions of two million passenger cars.

Dairy

Cows in dairy farms contribute to methane emissions during digestion, releasing this potent greenhouse gas into the atmosphere. Deforestation, spurred by growing dairy demand, compounds the problem. You might be wondering what the connection between cows and trees is, and the answer lies in animal feed.

Soy, grown in Brazil, often plays a main part in bovine livestock diets, and the growing demand has led to mass deforestation in some regions of that country as more space is needed to grow it (Boren et al., 2021). According to the WWF (2019), "Forest loss and damage is the cause of around 10% of global warming."

Recognizing the connection between our world's forests and seemingly unrelated things like dairy production underscores the links between everyday human consumption-driven activities and climate change. Implementing sustainable farming practices, refining cattle diets, and exploring alternative protein sources are crucial steps in addressing dairy farming's contribution to greenhouse gas emissions.

How Global Warming Happens

Far from being just a buzzword, "global warming" may very well come to be known as the most important phrase of the modern age. But what exactly does it mean?

Global Warming Defined

At its core, global warming is the mechanism by which our planet's temperature rises. It has the potential to set into motion a whole array of consequences that threaten to touch upon every corner of our lives.

Causes

When considering the forces that drive global warming, one glaring factor takes center stage: The release of greenhouse gasses. The human-caused emission of these gasses, which we've already discussed, has long been identified as the primary driver behind global warming.

Extreme Weather

The intensification of global warming elevates the increased likelihood and severity of extreme weather events. Heatwaves are likely to continue scorching many of the regions we inhabit, devastating hurricanes will continue to hit our coastal regions, spiraling with greater ferocity and bringing widespread damage and flooding, and prolonged periods of drought are likely to cripple ecosystems and communities. These types of extreme weather events, once anomalies, are set to become the new norm as the influence of global warming amplifies the power of nature's most formidable displays.

Other Effects

As global temperatures rise, the delicate balance of nature will be profoundly disrupted. Ecosystems will likely begin to falter as animal migratory patterns shift. Glaciers will continue their retreat, causing sea levels to rise. Global shifts in precipitation patterns could create a wide range of impacts, including devastating flooding and landslides. The delicate balance of life on Earth could be thrown into chaos by these changes, and the residual effects and fallout could extend to even the most remote corners of our planet.

The Daunting Task of Tackling These Threats

As the realities of global warming become increasingly clear, the magnitude of the challenges we must confront can feel overwhelming. The interwoven nature of human activities and their consequences on the environment necessitates comprehensive and concerted efforts on our part. We must intervene, but how?

From altering personal consumption patterns to advocating for systemic change, each action, no matter how small, plays a vital role in reshaping our trajectory. The road ahead will require not only individual responsibility but also collective collaboration, policy reforms, and technological innovations. While the enormity of the task may seem daunting, it's essential to recognize that every step taken toward sustainable living contributes to the broader endeavor of mitigating the impact of climate change.

What Is Climate Change?

Climate change encompasses the profound shifts that are happening in weather patterns and environmental conditions. As we further explore our shifting climate and what it means for our collective fate, we'll attempt to unpeel the various layers of these changes, peer into their origins, and process the far-reaching consequences they entail.

Climate Change Defined

Climate change isn't just a singular occurrence. Rather, it refers to the alteration of long-term weather patterns and global temperatures, driven by a combination of natural processes and, importantly, human activities. It extends beyond mere shifts in temperature, also factoring in things like rainfall increases or decreases, changes in sea levels, and the frequency of extreme weather events.

While climate variations have been a natural part of Earth's history, the rapid changes witnessed in recent decades, primarily attributed to human-induced factors, are a cause for concern. Understanding these dynamics is essential for grasping the severity of the challenges we face and the imperative for immediate action.

Quantifying the Impact Over Time

Quantifying the impact of climate change over time is a daunting task that requires meticulous data collection and analysis. Scientists and researchers rely on a wide array of methods to measure and interpret the changing climate trends, drawing from a vast array of data sources, including satellite monitoring, terrestrial weather stations, ice core sampling, and more.

By tracking indicators such as global temperature averages, sea-level rise, and shifts in ecosystems, scientific experts are able to paint a comprehensive picture of the scope and magnitude of the changes occurring on our planet. This data-driven approach is pivotal for assessing the urgency of our response and crafting effective strategies to mitigate the impact of climate change. It presents us with an undeniable reality: Climate change is real, and we must address it now!

THE THINGS WE DO THAT DRIVE CLIMATE CHANGE

As we've already touched upon, Earth's climate is not only influenced by natural processes but also by a broad range of human activities that release greenhouse gasses into the atmosphere. These gasses are emitted through activities such as energy production, natural resource exploitation, and industrial processes.

We've already taken a look at some of the lesser-known contributors to greenhouse gas emissions, so now it's time to take a closer look at the major players. Many of these activities have contributed to the release of greenhouse gasses for centuries, but that doesn't mean we can't turn things around by becoming more educated consumers and demanding change right now!

The Traditional Sources of Greenhouse Gasses

Burning Fossil Fuels

The energy we depend on for electricity, transportation, and heating worldwide still largely comes from fossil fuels like coal, oil, and natural gas–a staggering 80%, according to the Environmental and Energy Study Institute (2021). As these fuels burn, they release substantial amounts of CO_2 and other pollutants into the air, contributing significantly to global warming.

By moving toward green energy solutions, we can help wean ourselves off of dirty fossil fuel energy sources. Unfortunately, lobbying arms of oil companies and special interest groups, often funded by corporations with a vested interest in keeping the world reliant on dirty energy, have made the transition to green energy extremely difficult.

Deforestation

The clearing of forests for agriculture, logging, and urbanization reduces the planet's capacity to absorb CO_2. Trees play a crucial role in capturing carbon dioxide, and their removal disrupts this vital balance.

Tree replanting initiatives are one way to combat this. In the US, many states have mounted their own tree planting initiatives, and in Canada, the National Greening Program, which replants trees across the country, has been extremely successful, replanting 1 million trees last year alone (Tree Canada, 2023).

Industrial Agriculture

Industrialized farming practices, particularly livestock production, emit significant amounts of methane, an extremely potent greenhouse gas. As we already noted in our discussion of the dairy industry's impact on greenhouse gas emissions, cows and sheep release gas during digestion and also through the decomposition of their waste.

One way to reduce these emissions would be to start shifting our own dietary preferences. While this doesn't mean complete abandonment of meat consumption, embracing a diet that relies more on plant-based proteins reduces red meat consumption and can play a pivotal role in curbing methane emissions worldwide.

There have even been some recent initiatives to make lab-grown meat, and while the idea of manufactured meat may be off-putting to some, it's an important area of research and investigation and may play a role in feeding the world of the future.

Factories

Manufacturing and industrial activities release a mix of greenhouse gasses, which contribute to the overall warming effect. Fossil fuel combustion, often necessary to power the machines and generators that facilitate production processes, is a major culprit behind the release of carbon dioxide. Additionally, certain manufacturing processes and chemical reactions release nitrous oxide, further intensifying the warming effect.

The carbon footprint of the manufacturing industry isn't confined to individual factories or plants alone; it's the entire network of global manufacturing that's the real problem. The global nature of today's complex supply chains and logistics networks magnifies the carbon impact. Products often traverse the globe before reaching their final destinations, involving transportation via ships, planes, and trucks, each contributing its own share of greenhouse gas emissions.

One way we can start to mitigate the carbon impact of manufacturing is to put pressure on companies to start domestic manufacturing initiatives, reducing the carbon toll of shipping parts, components, and finished products between suppliers across the world. We can also play our own part by becoming more conscious consumers, spending less on products overall, or buying higher quality, even handmade or handcrafted items that we can use for decades to come.

Garbage

The decomposition of organic waste in landfills generates methane. However, proper waste management strategies can significantly reduce these emissions. The good news is that the waste industry has acknowledged the role of waste management in greenhouse gas emissions and is undergoing a transformative shift towards sustainability.

One example of innovative solutions in this area is carbon capture and utilization (CCU) technologies. These technologies capture emissions and make them into usable products like methanol, effectively recycling once-harmful atmospheric gasses into clean-burning, low-carbon-emitting fuels (*Carbon*

Capture and Utilization, 2023). In addition to these CCU methods, waste management companies are embracing circular economy principles to minimize methane emissions. Composting organic waste to produce nutrient-rich soil diverts biological waste from landfills where they'd normally just sit and emit gas and instead use it to grow food. In closed-loop systems, organic waste becomes a valuable resource, reducing the overall environmental impact.

THE CONSEQUENCES OF A WARMING PLANET

As the Earth's temperature steadily rises, the delicate fabric of life that spans across its ecosystems is being stretched to its limits, yielding a series of profound consequences that reverberate through nature and human societies. The effects of global warming are far-reaching and deeply interconnected, leaving an indelible mark on our planet.

The Growing Prevalence of Wildfires

Wildfires, intensified by escalating temperatures and changing weather patterns, have become an emblem of the planet's escalating climate crisis. The scorching heat waves of recent years have engulfed vast swathes of land in flames, leaving destruction in their wake. Unfortunately, kids growing up today are starting to consider these extreme wildfires the new norm. They don't react to the news on TV since they're used to it by now and witness it every summer.

In Greece–the place that initially inspired me to take a deep look into consumer options in everyday decisions—a summer of a record-breaking fire season is currently seeing blazes ravage landscapes around Athens. Out-of-control infernos like this have reached unprecedented proportions; in some ways, they're becoming commonplace, an incredibly frightening reality.

As these fires blaze through cities and towns, they're not only destroying natural ecosystems but also severely endangering human lives. As I sit here writing this book, I'm witnessing the horror play out in real-time, as Maui, Hawaii, has been thrown into turmoil by a wildfire, killing at least a hundred people, with over one thousand still missing (Arango & Healy, 2023).

Loss of Species and Habitat Disruption

A warming planet casts a menacing shadow over biodiversity, disrupting the delicate harmony of ecosystems that have thrived for millennia. Species face dwindling habitats as their homes transform due to shifting temperatures and altered landscapes. The impacts of climate change can cascade through the food chain, altering predator-prey relationships and destabilizing the balance of nature. The loss of species, whether big or small, can have a detrimental effect on the overall health of biological life itself. Even the absence of one insect species can throw off the balance of a natural ecosystem already teetering on the edge.

Weather Patterns

The changing climate threatens to severely throw our weather patterns out of whack, reshaping the moderate climate and conditions that have long made our world a mild, life-conducive environment. The sheer number of transformations emerging right now within the atmospheric dynamics of the Earth is what's most concerning. The compounded effects of heatwaves, droughts, rainfall, and all their residual impacts threaten both nature and human society at large.

Hotter, Longer Heatwaves

Heatwaves, once sporadic events, have now become recurrent, skin-searing events that are challenging human endurance and resilience. From Europe to the United States and beyond, the climbing thermostats witnessed in July 2023 alone have set new records, shattering historical norms (PBS, 2023). According to the World Weather Attribution network, the extreme heat that was endured in Europe, the US, and Mexico during the summer of 2021 would have been "virtually impossible" without human-caused climate change (Taylor, 2021a).

These prolonged periods of extreme heat we've seen in recent years have strained healthcare systems, caused energy demand spikes, and exposed vulnerabilities in urban planning, dispro-portionately affecting the elderly and marginalized communities.

Longer Periods of Drought

Increasingly frequent and severe droughts threaten to cripple ecosystems, agriculture, and communities worldwide. Regions that once relied on predictable rainfall patterns now grapple with extended dry spells, causing water scarcity and food inse-curity. The struggle for water resources intensifies, and with it comes the potential for conflict over these vital lifelines.

Vital sources of water are under threat in many regions of the world. The Colorado River, which has traditionally served as a vital lifeline for the entire Western and Southwestern United States, as well as a large swath of Mexico, has lost 10 trillion gallons of water since the year 2000 due to climate change (Milman, 2023). The tributaries of the river provide water to around 40 million people in the US alone (Siegler, 2021). Those further downstream will be the first to feel the heavy effects of this crisis, which has already garnered an official shortage declaration for the first time in history.

More Extreme Rainfall

Extreme rainfall events have surged in frequency and intensity, triggering catastrophic floods that leave devastation in their wake. From Pakistan to East Africa, deluges of water engulf landscapes, causing destruction, displacement, and loss of life. Floods undermine infrastructure, erode soil, and contribute to the spread of waterborne diseases, particularly impacting vulnerable populations.

As we discussed in highlighting the methane output of rice cultivation, whether in intentionally flooded or naturally inundated areas, the more of these types of flooded, anaerobic areas we have across the globe, the more methane-producing bacteria will thrive. A flooded world will mean more methane gas emissions, further compounding the problem.

Reduced Crop Yields

A warming climate poses a significant threat to global food security. Rising temperatures and shifting precipitation patterns disrupt agricultural cycles, leading to reduced crop yields. Corn and wheat, staple crops critical to global food systems, are particularly vulnerable. Decreased productivity threatens livelihoods, exacerbates hunger, and heightens the risk of societal instability in regions dependent on agriculture.

Vulnerable grain crops that are already under threat rely on massive amounts of water, but as rivers and tributaries dry up, water may have to be trucked in to keep farms going, furthering their carbon toll. Aside from irrigation systems being under threat, a changing climate could bring new diseases, parasites, and invasive species that could make agricultural production impossible in regions previously conducive to the growing of crops.

Health Challenges and Migration

As temperatures climb and ecosystems evolve, health challenges emerge, affecting both agricultural workers and livestock. Heat stress endangers those who toil in fields and farms while changing disease vectors and increased pests threaten livestock and crops alike. The intertwined relationship between human health and agricultural vitality underscores the consequences of a warming planet. The mortality rate amongst agricultural workers is already staggeringly high when compared to other types of workers, clocking in at 20%, according to a report by the Environmental Defence Fund and La Isla Network (2023).

According to that same report, today's average U.S. farmworker is subjected to "21 working days in the summer growing season that are unsafe due to heat." They also determined that these workers are poised to endure "an additional 18 days of work above safe heat levels if emissions peak before mid-century" and then start to decline. The prospects, as you might imagine, are even worse if worldwide emissions don't start to taper off by the mid-century (Environmental Defense Fund & La Isla Network, 2023).

MEASURING CLIMATE IMPACT

In making an effort to comprehend better and then begin mitigating the far-reaching consequences of climate change, understanding our carbon footprint stands as a pivotal first step. The good news is that doing this doesn't have to be such a daunting undertaking, as there are a number of easily accessible tools today that we can use to quantify the carbon impact of various activities.

Though there are a number of different carbon footprint metrics out there, each one providing different views and scopes of the impact of personal and business activities. Overall, the main goal is the same: To know your impact and hopefully do something about it! Before we can get into the nitty gritty of carbon footprint calculation, let's take a closer look at exactly what a carbon footprint is.

What Is a Carbon Footprint?

In basic terms, a carbon footprint is a convenient metric that helps us gauge the environmental impact of various human activities. This can include personal, household, or business activity. It aims to quantify the total amount of greenhouse gas emissions into a number we can easily understand and use to make more informed decisions.

Focusing specifically on CO2 and other equivalents, carbon footprint metrics aim to provide an understanding of all of the emissions stemming directly or indirectly from an individual, organization, product, or event. In essence, this figure helps show us the extent to which our personal choices, business decisions, and the processes that bring us the things we consume contribute to the unfolding climate crisis.

How Is It Calculated? And What Are Some Examples?

There are a number of different online tools available that can help you calculate your personal, household, or business-related carbon footprint, here are a few of them:

- **EPA Climate Calculator:** https://www3.epa.gov/carbon-footprint-calculator/
- **Conservation International:** https://footprint.conservation.org
- **CarbonFootprint.com:** https://www.carbonfootprint.com/calculator.aspx
- **UN Carbon Calculator:** https://offset.climateneutralnow.org/footprintcalc
- **UN Consumer Goods Calculator:** https://www.2030calculator.com/
- **UC Berkeley Calculator:** https://coolclimate.berkeley.edu/calculator

You can also type "carbon calculator" into your search engine of choice to get access to the latest, most popular tools.

Tracking CO2e Emissions

As we start to become familiar with the more advanced side of carbon accounting practices, the concept of carbon dioxide equivalent (CO2e) comes into play.

What Is the CO2e Metric?

The CO2e metric is a way of accounting for all the various greenhouse gasses, not just CO2. It converts the impact of other gasses, like methane and nitrous oxide, over a determined amount of time into a common unit of measurement, which is how much CO2 would have to be released to generate the same impact.

What Does CO2e Measure?

By incorporating all the greenhouse gasses and their impact into a singular metric, CO2e allows for a more complete assessment of our actual carbon footprint than looking at CO2 emissions alone.

What's the Difference Between CO2 and Co2e?

CO2 is a gas, and CO2e is a metric that comprises multiple greenhouse gasses and quantifies how they impact the Earth over time.

How Is CO2E Calculated?

Global Warming Potential (GWP) is a key factor in CO2e calculation. It quantifies how much heat a particular gas traps compared to CO2 over a specific time horizon. This allows for comparing the impact of different gasses in a standardized manner. The Intergovernmental Panel on Climate Change (IPCC) assigns GWPs for various gasses over different time periods, such as 20, 100, and 500 years.

The CO2e calculation involves a simple formula: Multiplying the amount of a greenhouse gas by its GWP. For instance, to calculate the CO2e of 1 ton of methane emissions over a 100-year time frame, you'd multiply the emission by methane's GWP for that specific time horizon.

Carbon Accounting

While powerful accounting and auditing tools, like carbon calculators, have aided in quantifying and managing our carbon footprint, there's a growing need for more qualified workers in the field of carbon accounting. Many of these new roles being carved out are based on the meticulous process of measuring, tracking, and reporting greenhouse gas emissions and providing a comprehensive overview of a business entity's environmental impact. As the global community accelerates efforts to combat climate change, carbon accounting is rapidly evolving into an important discipline. If you're the number-crunching type, it might even be for you!

Carbon Accounting Explained

Governments, businesses, and nongovernmental organizations (NGOs) alike are turning to carbon accounting practices to better comprehend and curtail their CO2e emissions. By deciphering complicated emission patterns and assessing environmental risks, carbon accounting practices empower decision-makers to formulate targeted strategies for emission reduction and resource optimization.

The demand for professionals skilled in carbon accounting methodologies will only increase as more governments and industry consortiums worldwide make regulations and recommendations that necessitate transparent carbon reporting from private companies. In the corporate world, sustainability and environmental managers are already embracing carbon accounting practices as an essential part of corporate social responsibility (CSR), aligning with eco-conscious consumers' and investors' expectations and helping foster brand integrity in a world where, fortunately, carbon impact is starting to be viewed as important.

Even King Charles has made a recent job posting for a sustainability data reporting manager at Buckingham Palace (Ibrahim, 2023), showing that the most stodgy old-world institutions recognize the urgency to incorporate sustainable practices into their operations.

Why Is It Important?

The relevance of good carbon accounting practices in the fight against climate change can not be understated. By accurately tracking CO2e emissions, we're able to clearly see the cumulative effect of our lifestyle, actions, and business activity on global warming and climate change. This data is the cornerstone of crafting effective mitigation strategies, policies, and initiatives that target the very heart of the crisis. As the climate crisis accelerates, the urgency of carbon accounting grows, elevating its role as a compass that can guide us toward a more sustainable future.

FOSSIL FUEL EXTRACTION AND EMISSIONS

The world's relentless reliance on fossil fuels has been a cornerstone of modern industrialization and economic progress. Yet, as we peer deeper into the impact of these worst climate change offenders, the complex picture and carbon toll of extraction and fossil fuel emissions and their direct consequences becomes even more glaring. Understanding the complex interplay between these elements and their real impact is critical when we start to confront the stark realities of our still fossil-fuel-reliant world and begin to consider alternatives that'll help us break away from these chains and shape a more sustainable future.

Understanding Fossil Fuels' Contribution to CO2e Emissions

Fossil fuel combustion releases a range of greenhouse gasses, chief among them carbon dioxide (CO_2) and methane (CH_4). Combustion of these fuels takes place across a broad range of human activities. Here are the most guilty offenders:

- **Transportation:** Cars, trucks, buses, airplanes, and ships rely heavily on fossil fuels like gasoline, diesel, and jet fuel for propulsion, contributing to a significant amount of worldwide CO_2 emissions.
- **Power:** Many power plants, especially those burning coal and natural gas, produce electricity by combusting

fossil fuels. These plants are a major source of emissions.

- **Heating and cooling:** Fossil fuels are commonly used for heating buildings and homes. Oil and natural gas are often burned in furnaces, which let out emissions through their exhaust.
- **Industrial processes:** All sorts of industries rely on fossil fuels for production processes, leading to emissions from industrial and manufacturing activities.
- **Residential and commercial energy use:** Beyond heating, fossil fuels like natural gas are also commonly used for cooking, hot water heating, and furnaces, amongst other energy needs in residential and commercial settings.
- **Agriculture:** Global agricultural activities, especially ones that rely on diesel-powered machinery for tilling, planting, and harvesting, rely heavily on fossil fuel combustion.

What Are Fossil Fuels?

Fossil fuels–coal, oil, and natural gas–are made of decayed prehistoric plant and animal matter, transformed by heat and pressure over millions of years. They are, unfortunately, still the energy sources most commonly relied upon today, powering everything from transportation infrastructure to energy sources in many places across the globe.

How Are These Fuels Formed?

Coal, which is mostly made of carbon, forms from decayed plant matter in the ground. Oil (petroleum) comes from ancient decayed biological life that once filled the oceans of the early Earth. Natural gas emanates from fossilized remains buried deep within the Earth. Extraction processes involve mining, drilling, and sometimes hydrofracturing processes, followed by a refining process necessary to transform crude oil into usable fuels.

The Clear Disadvantages of Fossil Fuels

The same attributes that make fossil fuels energy-dense and relatively cheap also make them environmental threats. Their combustion releases massive quantities of CO2, contributing to global warming and air pollution. The various extraction processes used to mine these fuels, particularly fracking, can devastate ecosystems and natural habitats.

The Consequences of Fossil Fuel Production

From extraction processes that scar landscapes and pollute water bodies to carbon-producing transportation networks that traverse continents, the production of fossil fuels takes a heavy toll on the environment. The emissions resulting from their combustion only further the ecological disruptions.

Mining for coal or drilling for oil disrupts ecosystems, upsets natural habitats, and generates a substantial amount of waste. Extracting fossil fuels through environment-damaging methods like fracking poses major threats to society.

What Is Fracking?

Fracking, short for hydraulic fracturing, is a technique used to extract natural gas and oil from underground rock formations. It's done by injecting a slurry made of chemicals and abrasives mixed with water at high pressure into the rock. This fractures the rock and releases the trapped fossil fuels.

The process has raised numerous concerns due to its potential environmental impacts, including water contamination and the release of methane during the extraction process. It can affect local and regional groundwater quality and can even lead to earth tremors as it blasts away rock formations, leaving underground tunnels that can destabilize the land above and can even exacerbate existing fault lines in already earthquake-prone areas.

Transporting These Fuels

The global supply chain for fossil fuels relies on vast fleets of ships, trucks, and pipelines. These transportation networks can lead to oil spills, leaks, and environmental contamination, widening the ecological toll of fossil fuel use.

Burning Fossil Fuels

The combustion of fossil fuels powers global economies while simultaneously spewing CO_2 and other pollutants into the atmosphere, intensifying climate change and compromising air quality.

Waste

Fossil fuel production yields massive amounts of waste, including slag heaps from coal mining and tailings ponds from oil extraction. These waste sites can lead to environmental degradation and water contamination.

The Clear Link Between Fossil Fuels and Climate Change

The undeniable connection between fossil fuel emissions and climate change is backed up by current-day scientific consensus. As the global economy continues to burn fossil fuels at unprecedented rates, the repercussions are manifested in the form of melting glaciers, rising sea levels, and intensified extreme weather events.

Revealing the dirty (not so secret) secrets behind fossil fuel extraction and emissions can serve as a clarion call for transformation. The path toward more sustainable energy sources beckons us further as we choke on each dirty spew from our remaining sources of fossil energy. The time is now to stand up and demand further innovation, cooperation, and an unwavering commitment to creating a healthier, more resilient planet.

ESCAPE FROM THE INFERNO: THE RHODES WILDFIRES

As I sit here right now, amid the picturesque landscapes of the idyllic Greek island of Rhodes, writing this chapter, a summer getaway has turned into a nightmare for a group of UK tourists.

Imagine you're on vacation, enjoying yourself on the beach, and then, within moments, the air becomes thick with acrid smoke, and the once-clear vista of the Aegean Sea becomes obscured. Panic grips your heart as you realize that the flames are advancing rapidly, encircling your beach chair. Amidst all the chaos, families start scrambling to gather their belongings, sweeping their young children into their arms, and dragging their luggage behind them like anchors through the sand.

Through a haze of smoke and uncertainty, you and the other guests make your way through flaming narrow pathways and roadblocks, your eyes seeking any way out. The journey to safety becomes an agonizing race against time as flames lick at the nape of your neck.

The real-life survivors of this real-life incident just happened to bear witness to the ferocity of climate-fueled wildfires, a force that will indiscriminately alter lives and landscapes as they continue to happen more frequently. Their harrowing experience serves as a somber reminder of the ever-growing threat of these types of events directly caused by climate change.

As the embers of Rhodes settle, and bodies, many of them burnt beyond recognition, are beginning to be recovered from the fire in Maui that's also happened as I write this book, the truth emerges in the shadows of smoke and rising temperatures: Our planet is in peril!

Throughout this chapter, we've explored the interconnected factors that demand our attention. But the story doesn't end here. The unfolding drama of climate change intertwines with political dynamics, revealing the dynamics of global power balance and the critical role it plays in safeguarding the Earth's future. In Chapter 2, we'll be exploring the relationship between human decision-making processes, government policies, and their ramifications for the fragile planet we all share.

2

POWER, POLITICS, AND OUR PLANET

Taking a closer look at the realities that climate change presents, a stark truth emerges: The path to a greener, more sustainable future is going to be largely paved through collaboration between governments and businesses. From the boardrooms of multinational corporations to the chambers of government, the influence of the decisions made and actions taken in these exclusive corners of power spreads far beyond their origins, leaving an indelible mark on the environment and the lives of people worldwide.

Consider this: A mere 100 companies have accounted for a staggering 71% of global greenhouse gas emissions since 1998 (Trendafilova, 2022). While each of us carries the weight of personal responsibility in our own actions and lifestyle choices, mega-conglomerates hold immense influence over the trajectory of climate change and have largely been able to avoid being held accountable for their actions. The unfortunate truth is that corporate interests, lobbying, and political donations can end up influencing regulatory and legislative actions. A complex web of power politics is spun this way in large, developed economies, with implications that play out on a global scale.

This chapter is dedicated to exploring the complex interplay between power, policy, and planetary health. We'll be taking a closer look at the ways in which the actions and policies of corporations and governments shape, perpetuate, and inform the climate crisis. While the concept of carbon accounting has gained traction within many large global corporations, and companies are beginning to take more responsibility for their CO2e footprint, these aren't the only metrics of environmental responsibility. Companies must also confront the other wasteful sources of emissions of the contractors and subcontractors they often rely on in their operations and supply chains. Additionally, they must take a closer look at things like product life cycle in order to reduce the staggering amount of consumer waste they contribute to.

Government intervention should be a cornerstone in the fight against climate change. In an era where individual efforts, though crucial, are never enough to turn the tide, the importance of collective, grassroots action becomes evident. Concerned citizens, along with the governments charged with representing their interests, wield the potential to drive comprehensive change through policies that protect ecosystems, support small eco-friendly farmers, makers, and service providers, promote green energy, and solve some of today's most pressing climate issues. The role of banks and the financial system in this narrative cannot be understated, as they are just now beginning to adopt sustainability objectives.

In this chapter, we'll take a look at a couple of corporate green initiatives that stand as beacons of hope amid the landscape of big business. The impact of such initiatives is starting to make waves, not only within these respective organizations but also across entire industries. Successful examples of these responsible practices offer insights into how they can reduce environmental impact while enhancing business reputation and maintaining popularity with consumers.

CORPORATE RESPONSIBILITY AND SUSTAINABLE PRACTICES

In an era of heightened environmental awareness, corporations are under increasing pressure to align their profit-seeking activities with sustainability principles. The transition to a greener, more responsible approach is essential—not only for the environment but also for the long-term viability of these businesses. From tracking carbon emissions to rethinking supply chains, companies are adopting concrete measures to mitigate their own environmental impact, representing a positive step forward. In this section, we'll take a look at a few different aspects of corporate responsibility and sustainable practices, shedding light on efforts currently being made to drive positive change.

Understanding the Carbon Footprint of Corporations

One of the key metrics used in evaluating a corporation's environmental impact is the "corporate carbon footprint." This comprises the total greenhouse gas emissions generated both directly and indirectly by an organization. Understanding this metric allows companies to gauge their contribution to climate change and identify areas for improvement within a broader scope than just looking at the impact of their CO_2e footprint. By analyzing emissions from various operational stages and supply chain activities, corporations gain insights into their carbon-intensive processes and can work towards reducing their overall environmental toll.

Quantifying the Corporate Carbon Footprint

Measuring the carbon footprint of a corporation involves quantifying the emissions associated with its activities. These emissions are categorized into three scopes.

What Are the Scope Emissions Rankings?

- **Scope 1 emissions:** Include direct emissions from sources owned or controlled by the company, like emissions from company vehicles, on-site power facilities or generators, and general operations.
- **Scope 2 emissions:** Include indirect emissions from electricity, heating, or cooling purchased from utility companies.
- **Scope 3 emissions:** Include all other indirect emissions along the value chain, including those from suppliers,

customers, and, critically, those accrued during the product life cycle.

Decarbonizing Operations, Supply Chains, and Business Models

To address their carbon footprints, some corporations are starting to embrace decarbonization strategies that extend far beyond their immediate operations. Exemplary companies are working to reduce emissions at every stage of their supply chain, from raw material sourcing to product disposal.

Some businesses are even gaining B Corp certification and reevaluating their entire business models to prioritize sustainability, exploring alternative energy sources, adopting circular economy principles, and designing products with a reduced environmental impact.

The Worst Offenders: Companies With the Biggest Carbon Footprints

Other corporations have continued their nefarious ways and have emerged as significant contributors to global greenhouse gas emissions due to the scale of their operations and supply chains. These worst offenders have attracted attention from climate action groups and other nonprofits and grassroots organizations for their substantial carbon footprints. According to the 2017 Carbon Majors Database report, these were the top 10 worst offenders, along with their respective percentage of greenhouse gas contribution:

- **China Coal:** 14.3%
- **Saudi Aramco:** 4.5%
- **Gazprom OAO:** 3.9%
- **National Iranian Oil Co:** 2.3%
- **ExxonMobil Corp:** 2.0%
- **Coal India:** 1.9%
- **Petróleos Mexicanos:** 1.9%
- **Russia Coal:** 1.9%
- **Royal Dutch Shell PLC:** 1.7%
- **China National Petroleum Corp:** 1.6% (Sustainability For All, 2017)

The Role of the Banking and Finance Sector

The impact of greenhouse gas-spewing corporations on the environment extends beyond their direct actions, encompassing their entire networks of customers and enablers. Who are these enablers, you might ask? Well, they not only include the companies engaged in business with them but also the banks that provide them with credit. While the focus on carbon-intensive companies is vital, it is equally important to scrutinize the financial support systems that allow these companies to operate.

In recent years, the financial sector has undergone a remarkable PR transformation as sustainability has taken center stage in corporate communications and messaging strategies. Banks, traditionally associated purely with profit-driven ventures and blind to the ecological implications of profit, are now beginning to integrate environmental considerations into their operations and investment arms.

This shift is driven by a growing recognition that their actions can significantly influence the trajectory of climate change. As the developed Western economies are starting to put a greater emphasis on the urgency of tackling climate issues, sustainability, and corporate responsibility is becoming a paramount concern for banks—at least in their rhetoric. Whether this gesture represents a genuine commitment or not is yet to be seen.

Some tangible first steps seen in recent years are gaining momentum. One such example is banks starting to offer improved financing terms to businesses that prioritize sustainability and demonstrate a commitment to reducing their environmental impact. Fully integrating sustainability principles into banking operations is going to involve multifaceted, multiforked strategies. These include embedding sustainability criteria into lending decisions, investment portfolios, and risk assessment processes.

Recognizing that individuals on this planet are keenly interested in aligning with an institution that promotes environmentally friendly endeavors, the global banking and finance system will likely become more incentivized to act as the rising tide of climate catastrophe continues.

In the world of mortgages and loans, climate risk assessment will continue to be a regulatory-driven and enforced norm. By assessing factors like rising sea levels, extreme weather events, and resource scarcity, banks and lenders can continue making informed lending decisions that account for climate-related risks. Climate risk assessment frameworks are one way that these ideas are being incorporated into the credit practices of banks and lenders. Such assessments allow banks to identify potential vulnerabilities in their lending portfolios and ensure that their exposure to climate-related risks remains manageable. This proactive approach helps protect both banks' financial stability and the broader economy from the potential shocks of climate change.

The global banking and finance sector's evolving role in upholding and introducing new sustainability practices underscores its significance as a potential enabler of change. As banks further embrace sustainability objectives, they're actively positioning themselves on the right side of the fight against climate change. By creating incentives for corporations to craft their own sustainability initiatives, banks are capable of helping cultivate a more resilient and greener economy that'll hopefully survive the shockwaves of future climate catastrophes.

Corporate Sustainability Initiatives

Aside from the emerging improvements in sustainability currently being enacted by banks, initiatives in many companies that depend on these banks for credit, banking, and financial services have themselves emerged as potent catalysts for positive change. Corporate initiatives continue to play a crucial role in driving companies toward more responsible practices that mitigate their environmental impact and hopefully contribute to a more sustainable future for their customers, as well as the global population at large.

Sustainability initiatives serve as vehicles for raising environmental awareness within organizations. By fostering a culture of consciousness about ecological challenges, these new programs and goals have the potential to empower employees at all levels to take collective action. Such initiatives not only educate employees about the importance of sustainable practices but also encourage them to integrate eco-friendly behaviors into their own daily routines.

One of the key advantages of these types of initiatives lies in their ability to mitigate the environmental impact of a company's operations and supply chain. By implementing energy-efficient technologies, optimizing resource consumption, and adopting circular economy principles, corporations can effectively reduce their carbon footprint. These goals extend beyond internal operations, potentially influencing the behavior of suppliers, partners, and stakeholders throughout the value chain.

Incorporating sustainable practices into corporate operations benefits the environment while also contributing to a company's long-term viability and the overall satisfaction of its employees and customers. Reducing greenhouse gas emissions, conserving resources, and minimizing waste not only align with global climate targets but also offer a tangible advantage in the marketplace. Forward-thinking organizations that prioritize sustainability attract environmentally conscious consumers, investors, and partners, resulting in enhanced brand reputation and financial performance.

The trend toward corporate sustainability is also bolstered by a shift in consumer preferences. As consumers become increasingly environmentally aware, they're beginning to demand more products and services that align with their values. Corporations that put sustainability at the forefront can tap into this growing market, gaining a competitive edge and fostering customer loyalty.

Big Climate Goals

There are several examples of corporate sustainability initiatives with big goals set. For example, America's original superstore, Walmart, has defined a path toward zero emissions by 2040, and their Project Gigaton initiative aims to reduce their carbon output substantially by 2030.

While these long-dated goals are commendable, it's yet to be seen if big box retailers are sincere or simply trying to appeal to public sentiment through greenwashing, a subject that we'll discuss later in this chapter. Walmart was recently hit with a $3 million Federal Trade Commission (FTC) penalty for misleading the public on the eco-friendliness of a bamboo-based rayon textile they were selling. Discount retailer Kohls was also implicated in the FTC proceedings for similar claims. Big-box retailer Target has set some of its own goals toward reducing its carbon emissions and adopting sustainable sourcing practices—but as with most of these companies, the tangible results are largely yet to be seen.

Across the pond in Europe, British media broadcast and telecommunications company SkyTV has taken steps to become carbon-neutral by 2030, emphasizing the importance of the media's role in addressing climate change. Swedish furniture manufacturer IKEA, long a global leader in sustainable practices, has committed to becoming a fully circular and climate-positive business by 2030.

Global initiatives like the United Nations' "Speed Up, Clean Up" campaign have encouraged businesses all over the world to accelerate their transition to clean energy and implement sustainable waste management practices. As more and more companies continue to clean up their acts, governments across the world are charged with the task of bringing in new emissions regulations, tax incentives, and a whole range of other legislation to address the climate crisis.

GOVERNMENT POLICIES AND CLIMATE ACTION

In the face of the monumental challenges posed by climate change, the importance of government action cannot be overstated. While individual efforts undoubtedly play a vital role, the sheer scale and complexity of the climate crisis require comprehensive policies and response plans at the highest levels of government.

The transformative changes necessary to tackle climate change represent a massive undertaking. The intricacies of global economies, industries, and infrastructures demand a broader approach involving systemic shifts driven by governments worldwide who must take action now! Government intervention is what will take a pivotal role in addressing the fallout from the global climate crisis.

The regulatory power, resources, and ability to implement large-scale policies make government agencies crucial agents of change in mitigating environmental degradation. Centralized governments have a unique position to influence industries, set regulations and emissions caps, and allocate resources to drive a sustainable transition.

How Can Governments Best Confront Climate Change?

To effectively confront climate change, governments have a range of responsibilities to tackle. These include:

- **Protecting and restoring key ecosystems:** Governments can implement policies to safeguard and restore vital natural ecosystems, such as forests, wetlands, and marine habitats. These remaining biodiverse areas play a critical role in carbon sequestration and preserving biodiversity.
- **Supporting small agricultural producers:** Policies that support small-scale and sustainable agriculture can enhance food security, reduce deforestation, and promote resilient farming practices.
- **Promoting green energy:** Governments can incentivize the transition to renewable energy sources such as solar, wind, and hydroelectric power. This shift reduces reliance on fossil fuels and curtails greenhouse gas emissions.
- **Combating short-lived climate pollutants:** Addressing air pollutants like methane and black carbon (soot) can yield rapid results in curbing warming and improving air quality.
- **Bet on adaptation, not just mitigation:** In addition to mitigation efforts, governments can develop strategies

to adapt to the impacts of climate change. This includes preparing for extreme weather events, sea-level rise, and changing agricultural conditions.

- **Providing detailed action plans:** Governments can create comprehensive action plans that outline targets, strategies, and specific measures to be implemented over defined timeframes. Clear accountability mechanisms ensure that progress is tracked and reported.
- **Bolder incentives and mandates:** World powers can enact stronger regulations and incentives that push industries and consumers toward sustainable practices, thereby accelerating the transition to a greener economy.
- **Boosting innovation through funding:** Investment in research, development, and innovation can drive the creation of new technologies and solutions that contribute to climate resilience.
- **Improving the design and delivery of green initiatives:** Governments can ensure that green initiatives are effectively designed, communicated, and accessible to all segments of society.
- **Serving as role models:** Administrations can lead by example, integrating sustainability into their operations, procurement, and infrastructure projects, setting a precedent for other sectors.
- **Promote a whole-of-society, people-centered approach:** Leaders can encourage citizen engagement, reach out to and mount partnerships with other governments, encourage participation in civic society,

and even influence the private sector to make people-centric changes to their business operations.

As you think about the things that are crucial for governments to address, think about how you could use these ideas to advocate for climate solutions within your own community. By urging governments to prioritize climate action, citizens like you and me around the world have the power to contribute to the formulation and implementation of policies that drive meaningful change, ensuring a more sustainable and secure future for all of us.

COLLABORATION FOR A GREENER FUTURE

As the world continues to grapple with the urgency of addressing climate change, it's clear that no single entity can tackle this challenge alone. The need for collaborative efforts between multiple stakeholders is more apparent than ever. This is where the power of partnerships and collective action comes to the forefront—driving innovative climate solutions and forging a path toward a sustainable world.

One aspect of this dynamic that's worth a closer look is the varying approaches different regions take to address climate change. For instance, when you look at the way the United States and the European Union address these pressing issues, the solutions vary. Not only do the policies differ, but the targets and strategies employed can also vary, sometimes drastically. While the specific implementations of climate action can vary between world regions, the common thread remains the significance of unified efforts that transcend borders for the common cause of human survival.

Cross-Collaboration Between Governments, Private Sector Companies, and NGOs

Cross-collaboration and cross-pollination of ideas is what's at the heart of addressing climate change, and this extends to partnerships between governments, private sector companies, and NGOs. The combined strength of loosely-banded grass-roots groups, along with more formalized organizations and institutions around the world, can lead to innovative solutions, widespread adoption of sustainable practices, and the creation of robust policies that drive change on a global scale.

The Vital Role of Grassroots Movements and Activism

Grassroots movements and community engagement are extremely important factors driving collaborative action. Local communities play a crucial role in driving environmental awareness and advocating for change at regional levels. These efforts, when connected to broader initiatives, have the power to strengthen their impact and drive meaningful shifts in behavior and policy.

Consumer Preferences: Driving Sustainable Change

Consumers also wield remarkable influence in shaping corporate behavior through their choices. The growing demand for sustainable products and services has prompted companies to reconsider their practices and offerings, helping bring about a culture of accountability and responsibility. Collectively, consumers have the power to drive corporate strategies toward greener alternatives while holding businesses accountable for their environmental impact.

The Washing Effect

An important part of making more involved, thought-out, and active consumer choices is staying vigilant in the face of deceptive corporate practices and misleading marketing and advertising campaigns. The practice of "greenwashing," where companies exaggerate or falsely advertise their environmental efforts, remains a valid concern. Educated conscientious consumers have to learn to be discerning about the intentions and actual impact of any green corporate initiatives they encounter. This can be achieved by staying informed about the real data surrounding marketing claims. By doing the research to see if companies' claims are true, we can easily distinguish genuine commitment to sustainable practices from mere marketing ploys.

The concept of "youthwashing" is also worth mentioning. It refers to companies using youth-driven imagery and language to appear environmentally responsible while failing to make any concrete, substantial changes. Being more aware of these tactics can also help us distinguish between genuine actions and superficial gestures of corporations adjusting their marketing messages to keep up with the times and capture the attention of a younger, more climate-informed consumer.

Private Sector Sustainability Case Studies

Highlighting and celebrating some of the more successful examples of corporate sustainability campaigns remains important. A few companies have enacted sustainability initiatives that have influenced consumer behavior and serve as shining examples.

In the apparel industry, successful examples include American outdoor clothing manufacturer Patagonia's circular economy "Worn Wear" program, which encourages customers to extend the life of their products through repair, reuse, and recycling, fostering a culture of longevity and reduced waste.

Overseas, world-famous Danish toy maker LEGO has shown its commitment to green manufacturing by becoming a World Wildlife Fund Climate Savers Partner. This strategic partnership showcases their dedication to reducing the carbon impact of their manufacturing by using environmentally friendly materials in making their popular toy plastic building bricks.

Collaboration Toward a Better Future

Dynamic collaboration between entities of all sizes is set to remain a cornerstone of the fight against climate change. From governments to grassroots activists, consumers to corporations, genuine strategic climate partnerships can drive systemic change and pave the way toward a greener future.

As we conclude this chapter, you can see ever more clearly that addressing the climate crisis requires multiforked, united efforts that span across economic sectors and geographic regions. We'll be further exploring this theme in Chapter 3 as we look at the interconnectedness of human society and endangered animal habitats, highlighting the pressing issues of environmental and climate justice.

3

BALANCING THE SCALES

The issues of climate change, biodiversity loss, and environmental justice are inherently tied together, forming a complex view of today's global realities. The urgency of addressing climate change has given rise to an awareness that goes beyond the realms of science and policy. It has coalesced into a poignant acknowledgment of the symbiotic nature of our planet's well-being, the fragile balance of its ecosystems, and the imperative of equity for all.

As we've already discussed a bit earlier in this book, there's a profound connection between climate change and biodiversity. The quest for climate justice and the fundamental need for a fair and equitable world can be seen as complementary issues to the climate problem itself. The voices of scientists, activists, and concerned communities have spoken: They're all in favor of making our best effort to save the myriad species that share our planet–including our own.

The impact of climate change is no longer confined to distant, unforgiving climates or relegated to being the concern of future generations, a highly problematic belief. The pressing reality we must confront is that the effects of this crisis aren't evenly distributed. Vulnerable communities, often least responsible for the causes of climate change, bear the brunt of its consequences. The disparities in impact are a stark reminder that the fight against climate change is intrinsically tied to the pursuit of equality, justice, and equity.

We can see this dynamic at play in the recent tragedy in Maui, the Hawaiian islands, a place with an already polemic colonial legacy, now neglected by politicians and emergency management agencies. This is happening not in some distant, remote, or unknown country we wouldn't associate with tourism, but in a place that today is visited by people all around the world due to its association with honeymoon and happiness. With 1,300 people still missing after the fire, we can almost feel the scorching heat and choke on the barely breathable air in sympathy with the people of Hawaii.

As we jump into further exploring the threats to biodiversity and our earth's ecosystems, we should still take a moment to recognize the beauty of the delicate dance of life that sustains our world. The staggering variety of species, each with its unique role, paints a picture of interconnectedness that underscores the importance of safeguarding what we have. For species great and small, the implications of decisions humans make today are profound, their very survival hinging on our ability to maintain and preserve our delicate planet for future generations.

As the fury of extreme heat disrupts the harmony of forests and the devastating impact of ocean acidification takes its toll on marine life, stories of species on the brink of extinction can serve as a potent reminder of the fragility of life. Amidst the challenges we face in confronting these issues, there are stories of hope. Conservationists work tirelessly to protect ecosystems, restore habitats, and advocate for better solutions. The concept of climate justice emerges as a guiding light, urging us to address not only the environmental but also the social dimensions of the crisis.

When climate refugees and internally displaced persons (IDP) tell their stories, listen. Those displaced by the wave of climate chaos that has already hit us help us better understand the human impact of climate change. Through listening to their lived experiences, we become better poised to confront the reality of the climate crisis as a global issue that demands global solutions.

ECOSYSTEMS AT RISK

Our planet is graced with an astonishing array of biodiversity. Nature has blessed us with a wide range of species that collectively comprise the ecosystems of diverse climates across the globe. Biodiversity, the variety of life that our survival depends upon, is the very foundation upon which our world's ecosystems are built. Part of becoming more climate-conscious is being more aware of ecosystems at risk and gaining a greater understanding of the multifaceted relationship between biodiversity and the environments they inhabit.

What Is Biodiversity, and Why Is It Important?

Biodiversity encompasses the dazzling array of diversity of life found on our planet, from the towering trees of ancient forests to the microscopic organisms thriving in the depths of our oceans. It's the governing principle that forms the backbone of functional, healthy ecosystems. The importance of biodiversity is multifaceted, affecting many parts of our own human existence.

The diversity of species is what gives ecosystems resilience. The presence of a variety of flora and fauna within an ecosystem provides a safety net against outside disturbances and threats. Each species contributes its unique abilities, ensuring the system's capacity to recover from environmental changes and weather events. Biodiversity bolsters ecosystems as well as human settlements within proximity by providing air purification, groundwater water filtration, and pollination of crops that sustain human life.

The Various Kinds of Biodiversity and the Importance of Maintaining Symbiosis

Biodiversity can be categorized into three main types: genetic, species, and ecosystem diversity. Genetic diversity within a species ensures adaptability, while species diversity presents us with a vast array of unique-looking creatures. The concept of ecosystem diversity as a whole is based around maintaining the distinctive character of various habitats, each hosting a unique cast of biological life whose life cycles play out upon its stage.

Ecological Niches

Ecosystems can be considered as networks of relationships and dependencies, where every species has a role to play—a niche to fill. Ecological niches denote the unique roles that each species serves within an ecosystem. The harmony of ecosystems depends on species largely sticking to their niches.

The loss of even a single species in a dynamic ecosystem can lead to a cascade of consequences, disrupting the equilibrium and functionality of large areas of undisturbed land. The interconnectedness of species serves as a poignant reminder that every loss, every climate-fueled migration, dispersal, or extinction, no matter how seemingly inconspicuous, reverberates through the natural world.

The Effects of Extreme Heat on Forest Ecosystems

The threats of rising temperatures, prolonged heat waves, and wildfires are casting an ominous shadow over the world's forests. Extreme heat has become a recurring theme with the frequency and severity of wildfires surging, leaving a trail of devastation in their wake. The very ecosystems that were once the cradle of life are now subjected to unimaginable climate-fueled destruction.

Forest fires, while often employed in a controlled manner for forest management, can unleash havoc upon plant and animal life when they get out of control. These fires can transform once-lush landscapes into barren, scorched earth. The aftermath of these infernos is characterized by habitats being stripped of their biodiversity. Heat stress stifles tree growth, affects seed germination, and tips the scales against vulnerable animal species struggling to survive.

The Rising Tepid Tide: How Climbing Temperatures Worldwide Are Disrupting Oceanic Ecosystems

The world's oceans, vast and seemingly boundless, are not immune to the far-reaching impacts of climate change. Ocean acidification, an often-overlooked consequence, is a silent yet serious threat to marine life. This phenomenon occurs as excess carbon dioxide in the atmosphere dissolves into seawater, altering its chemical composition and acidity.

Ocean acidification presents an ominous challenge, disrupting marine ecosystems from their very foundations. As the oceans become more acidic, coral reefs, mollusks, and a plethora of marine organisms face the risk of dissolution and decay. The fragile balance that sustains marine biodiversity is thrown into disarray, threatening the stability of fisheries and the communities that depend on them.

Ocean Acidification: What Is It Caused by, and Why Is It a Problem?

As carbon dioxide levels rise in the Earth's atmosphere, the oceans absorb a significant portion, altering their chemistry. The heart of the issue lies in oceans' balance between carbon dioxide and carbonic acid. As more carbon dioxide dissolves in the seawater, carbonic acid forms, leading to an increase in the concentration of hydrogen ions. This contributes to a decrease in ocean pH levels. As pH levels decrease, the oceans become more acidic, disrupting the delicate balance upon which marine life depends.

The consequences of ocean acidification can reverberate through all types of marine ecosystems. The primary concern is the impact it can have on marine organisms that rely on calcium carbonate to build their shells and skeletons. As oceans become more acidic, the availability of carbonate ions diminishes. This scarcity makes it increasingly difficult for species like corals, mollusks, and shellfish to form and maintain their protective shells and structures.

Declining Marine Biodiversity

Aside from ocean acidification's effects on mollusks, it poses a grave threat to marine biodiversity at large, as it can attack the very ecosystems they inhabit. Coral reefs, often referred to as the rainforests of the sea, epitomize this vulnerability. These living ecosystems, home to a breathtaking array of marine life, are built upon the calcium carbonate structures of coral polyps. As ocean acidification erodes these structures, the foundation of coral reefs can erode, leaving countless plant and animal species without a habitat.

Furthermore, the disruption extends to the food chains that sustain marine ecosystems. From zooplankton to larger predators, the health of marine ecosystems depends on predictable, steady symbioses. When certain species falter due to the impacts of ocean acidification, the collateral effects can disrupt the relationships and interdependencies between species, rapidly spreading through the ecosystem and threatening the stability of entire regions.

Effects on Fisheries and Coastal Communities

The ramifications of ocean acidification extend beyond marine ecology alone, as they also affect coastal communities. Many communities around the world depend on fisheries for sustenance and to make their livelihoods. As ocean acidification disrupts food chains, the abundance and diversity of fish populations decline. This diminished fish supply threatens fishermen's livelihoods and also the availability of a vital low-carbon impact protein source for millions.

Coastal communities, often among the most vulnerable to the effects of climate change, bear the brunt of the repercussions. As fish populations decline, the economic well-being of these communities is undermined. Additionally, the loss of vital fish populations places increased pressure on already vulnerable food security systems in some coastal regions and island nations of the world.

Migration Pattern Disruption

Climate change is a driver of animal migration, one which threatens to alter the well-worn seasonal travel paths of various species. The rhythms and flows of migration, essential for many species' survival, get thrown into disarray by the changing climate. As temperatures shift and habitats transform, species are compelled to change their historical migration habits.

Migratory patterns are closely intertwined with the availability of resources, making them a barometer of ecological change. As temperatures rise, certain species are forced to take off on their seasonal migratory paths earlier than usual. Others may seek refuge in cooler climates, abandoning habitats that can no longer sustain them, often landing them in foreign, unfamiliar ecosystems where their presence can throw off the balance of other local species or lead to the displaced animal becoming easy prey.

Across the planet, ecosystems at risk of migratory shift are a harbinger of the dynamic interplay between biodiversity and environmental rhythms. It's important to remember it's not all doom and gloom, however; as plant and animal life further faces the strains of climate change, signs of adaptation and resilience are ever-present, painting a vivid portrait of a world in flux, and on the brink of a great shift that many will not survive.

CLIMATE CHANGE AND SPECIES EXTINCTION

Five mass extinctions have already shaped the course of life on our planet, leaving behind their indelible marks on the fossil record. Yet, today, as humanity treads upon the threshold of a new era, a chilling phenomenon is unfolding—the sixth mass extinction. Unlike its predecessors, this coming extinction event is driven not by cosmic cataclysms or natural upheavals but by our own actions, particularly the relentless assault against the Earth at the hand of human-induced climate change.

The Sixth Mass Extinction

Scientific consensus is that there have been five extinctions in the past 500 million years, the last one being 65 million years ago and the most devastating being the Permian mass extinction 250 million years ago (Begum, 2021).

This cataclysmic extinction is also referred to as The Great Dying because it wiped out more than 95% of all species (Begum, 2021). The sixth mass extinction now looms as a dire consequence of the complex range of challenges posed by human activities on Earth.

As temperatures climb, habitats vanish, and ecosystems are disrupted, causing species across the globe to grapple with unprecedented conditions. To better understand the urgency of this crisis, we have to look at its drivers, process its implications, and then rally to confront it head-on!

What's Propelling Us Toward the Sixth Mass Extinction?

At the heart of this ecological upheaval is climate change. As repeatedly emphasized throughout this book, climate change alters ecosystems, accelerates habitat loss, and amplifies the threats faced by countless species. Human actions, from deforestation to greenhouse gas emissions, have set the stage for a cascade of extinction events that will comprise the sixth mass extinction.

Why Should We Care About It?

This impending mass extinction isn't a distant abstraction relegated to academia and scientific circles; it's a real, coming crisis that we must confront. The specter of extinction is a deep one to process, as it means that we will cease to exist.

The Consequences of Taking No Preventative Action

The magnitude of the crisis demands urgent action, as the consequences of inaction are profound and far-reaching. Our world faces a cascade of interconnected catastrophes that threaten to reshape life as we know it.

Biodiversity loss will jeopardize global food security, exacerbating food shortages and impacting agricultural yields. The connections between species and ecosystems, essential for disease control, pollination, and water purification, will begin to fray. Eco-tourism, often centered around enjoying the beauty of diverse animal species and diverse landscapes, will disappear as habitats degrade and species vanish.

The consequences of inaction go beyond the practical. The loss of human life erases millions of years of evolution and extinguishes the unique stories we carry with us. Cultural, spiritual, and aesthetic values interwoven with nature's diversity are imperiled. The loss of humankind closes doors to unexplored realms of scientific discovery, limiting our understanding of our own planet and our vast universe.

The sixth mass extinction is not an isolated event; it's a call to action that echoes through time. The choices we make today hold the power to tip the balance toward a future teeming with life or one marred by scarcity, instability, and inevitable death.

As we stand at the crossroads of history, the urgency to address this extinction is resounding. The echoes of extinctions past serve as reminders that once a species is lost, it cannot be resurrected. Whereas past extinctions were inevitable, we still have the power to make the changes necessary to avoid this one.

The Emerging Biodiversity Crisis: What Happens When Current Extinction Levels Surpass the Natural Background Rate?

The wheels for the coming extinction have been set into motion, and it's currently unfolding before our eyes. This is all happening against the backdrop of escalating extinction rates. The natural background rate is the rate at which species would naturally go extinct in the absence of human influence. Historically, this rate has hovered around one to five species per year (Ferguson, 2019), a gentle rhythm of nature's ebb and flow. However, in our era, the pace is becoming worrying.

Shocking statistics reveal that species are vanishing at an alarming pace—1,000 to 10,000 times the background rate (Editors of Encyclopaedia Britannica, 2020). This unprecedented surge places our planet in the throes of a biodiversity crisis, where the diversity of life that has thrived for millennia now teeters on the brink of oblivion. Essential pollinators such as honey bees are under threat, demonstrated by their managed colony loss in the US spiking to over 60% in 2022 (Bee Informed, 2023).

Animals That Might be Extinct by 2100

- **Rhinoceros:** Rhinos, such as the Sumatran and Javan rhinos, are threatened by poaching practices that strip them of their horns and habitat loss due to human activities. Conservation efforts are crucial to protect these majestic creatures.

- **Saola:** The saola, also known as the Asian unicorn, is one of the world's rarest mammals. It's a critically endangered species found in the forests of Vietnam and Laos. Habitat loss and illegal hunting are driving factors for its decline.
- **Cat Ba langur:** Found on Cat Ba Island in Vietnam, the Cat Ba langur is one of the rarest primate species. Habitat loss, fragmentation, and hunting have drastically reduced their numbers.
- **Vaquita dolphin:** The vaquita is the world's most endangered marine mammal, residing in the Gulf of California. It faces extinction due to bycatch in illegal fishing operations for another critically endangered species, the totoaba fish.
- **Emperor penguin:** Inhabitants of Antarctica, emperor penguins are highly threatened by the impacts of climate change. Melting ice and shifting food sources can affect their breeding and survival.
- **Bornean orangutan:** The Bornean orangutan's survival is threatened by habitat loss due to deforestation and palm oil plantations. They're critically endangered, and their populations are rapidly declining.
- **Amur leopard:** The Amur leopard is one of the most endangered big cats, with less than 100 individuals left in the wild. Habitat loss, poaching, and human encroachment on their natural habitat are major threats.

- **Sumatran elephant:** Habitat loss and fragmentation, as well as conflicts with humans, are endangering the Sumatran elephant. They are also threatened by poaching for their ivory.
- **Sumatran tiger:** The smallest tiger subspecies, the Sumatran tiger, is critically endangered due to habitat loss and poaching. Their population has declined significantly.
- **Finless dolphin:** These dolphins are threatened by habitat degradation, pollution, and bycatch in fishing gear. Their unique appearance, with a lack of dorsal fins, makes them susceptible to entanglement.

My Trip to the Galapagos

My travels to the Galapagos Islands a few years ago with my children remain etched in my memory as an experience that touched my heart and opened my eyes to the fragility of our planet's biodiversity. As we explored the Galapagos, I was struck by the chance to witness rare and incredible animals in their natural, untouched environment.

Amidst all the awe-inspiring beauty I witnessed, which included giant turtles, huge lizards of various colors, and birds almost as big as my eight-year-old daughter, I felt an underlying concern that tugged at my heart. These islands, which are a living laboratory of evolution and biodiversity, are facing the same challenges that threaten countless ecosystems worldwide. The delicate balance of life is at risk, as many of these animals are on the brink of extinction.

My Galapagos journey was a humbling reminder of the responsibility to protect and preserve the incredible diversity that enriches our planet. The very creatures that fascinated my family with their uniqueness are the ones we must work tirelessly to safeguard. Our human impact on the environment, through climate change and habitat destruction, reverberates across these islands and far beyond.

I also thought about how lucky my kids were to be taken on such a trip and to have the opportunity to spend two weeks in a small eco-travel group with a marine biologist who would explain every single detail. Most kids will never have this exposure because this is the only place in the world where you can see things like this, and it's just too far and too difficult to get to.

Standing on the shores of the Galapagos, I realized that our collective efforts matter. Our choices, actions, and advocacy have the power to make a difference in the fate of these unique animals and the ecosystems they inhabit. The Galapagos Islands are not only a treasure trove of rare creatures but also a clarion call to safeguard the world's biodiversity, ensuring that generations to come can marvel at the beauty and diversity that enrich our lives.

CONSERVING BIODIVERSITY AMIDST A CHANGING CLIMATE

Amidst the challenges posed by a changing climate, the importance of safeguarding our planet's biodiversity has become more pressing than ever. Establishing and maintaining protected areas is a crucial strategy in the face of environmental threats. These protected havens, including state and national parks, wildlife preserves, and oceanic sanctuaries, serve as vital refuges for vulnerable species and their habitats, offering a chance for survival and adaptation to the extreme stress that climate change brings.

The Importance of Protected Areas and Natural Habitats

National parks and state parks are often touted by eco-lovers as the crown jewels of national and state conservation efforts. They play a pivotal role in preserving the diversity of life as the impact of our civilizations encroach on these spaces. Also, most importantly, perhaps, they provide areas where native flora and fauna can thrive. These protected areas not only harbor endangered species but also act as natural laboratories, allowing researchers to study ecosystems and gain insights into how they function and adapt to climate change.

Wildlife reserves and sanctuaries offer similar benefits, serving as places of respite for species that have been pushed to the brink of extinction. These havens are carefully managed to mimic the creatures' natural habitats, giving them a chance to recover and eventually be reintroduced into the wild. Marine sanctuaries, meanwhile, contribute to the preservation of marine biodiversity, providing shelter for various species from the threats of overfishing, pollution, and habitat destruction.

Ecosystem Restoration Initiatives

Ecosystem restoration and rewilding campaigns are innovative strategies intended to help rehabilitate degraded habitats and revitalize struggling ecosystems. Restoration focuses on ecosystems that have been damaged or destroyed, with the ultimate goal of returning them to their natural state. Rewilding, on the other hand, focuses on reintroducing native species into their former habitats and then sitting back while allowing natural processes to shape the environment.

On Rewilding

Rewilding is not a substitute for conservation; rather, it complements the efforts to protect and preserve existing habitats. It's an approach that acknowledges the intricate connections between species and their environments, aiming to restore ecological balance and revive natural processes and symbioses. By reintroducing species that have fled due to human activities, rewilding seeks to mend the natural processes and cycles that have been disrupted.

Is Rewilding a Necessary or Appropriate Solution?

Rewilding holds promise in a variety of landscapes and ecosystems around the world, particularly in areas where human activities have disrupted and driven species to the brink of extinction. Regions that have suffered habitat loss, degradation, or fragmentation due to activities like deforestation, urbanization, or industrialization can also benefit from these types of initiatives.

The Benefits

Rewilding efforts can be useful in more fragmented landscapes, where wildlife corridors are reestablished to connect isolated habitats, with the intent of allowing for the movement of species. It can also be impactful in areas where apex predators have been eliminated, leading to imbalances in the ecosystem. The reintroduction of predators to these habitats can have cascading effects that help regulate prey populations and restore ecological dynamics.

Sustainable Land-Use Practices

Sustainable land-use practices can also play a vital role in supporting biodiversity conservation. Land management methods that use natural resources economically while treading lightly on them are a great way to meet the current demand for crops without compromising the integrity of the land for future use.

Sustainable Land Management Methods

Methods such as agroforestry, terracing, and sustainable forestry aim to improve the relationships between human agricultural activities and the natural environment.

- **Agroforestry:** Agroforestry combines agriculture and forestry, where trees or woody plants are intentionally integrated into agricultural systems. This practice enhances ecological diversity, improves soil fertility, and boosts ecosystem resilience against climate change. In agroforestry systems, diverse crops, ranging from annual crops to fruit trees, are planted alongside trees. This diversity on farm fields helps prevent soil erosion, enhances nutrient cycling, and provides shade for crops, contributing to higher yields and healthier natural landscapes.

- **Terracing:** Terracing is an ancient land management practice that transforms steep slopes into level terraces, creating arable areas while also controlling soil erosion and water runoff. This method is especially useful in regions with hilly terrain already prone to degradation due to erosion. Terraced landscapes can increase land productivity and help stabilize ecosystems while protecting downstream areas from sedimentation.
- **Sustainable forestry:** Sustainable forestry is an approach that seeks to maintain the long-term health and vitality of forest ecosystems where logging industries are active. This method balances the need for lumber with the ecological support functions of forests, such as carbon sequestration and providing habitats for animals. Sustainable forestry involves practices such as selective logging, where only mature trees are harvested, leaving younger trees to keep growing. It also emphasizes reforestation efforts to replenish what's harvested.

The Benefits

These sustainable land management methods share a common thread: Recognizing that human land use doesn't have to always come at the expense of the environment. Rather, they demonstrate that human activities can actually be transformed in ways that enhance ecological resilience, conserve biodiversity, and help mitigate the impacts of climate change.

By adopting more practices like these, we can help preserve the natural world, keeping it in good shape for future generations.

Through careful integration of sustainable practices, we can create landscapes that flourish with ecological diversity while also fulfilling vital human needs and resources like food and building materials.

Case Studies

Case studies from around the world offer compelling evidence of the positive impact of these practices.

- **Tree planting in Niger** has transformed barren landscapes into thriving forests, revitalizing land and helping the communities that depend on it.
- **Terracing in China** has helped curb soil erosion and improved agricultural productivity.
- **Sustainable forestry in Costa Rica** has demonstrated that responsible logging can coexist alongside conservation efforts.
- **Agroforestry in Kenya** has shown that agricultural practices can enhance biodiversity while also providing sustainable livelihoods.

ENVIRONMENTAL RACISM AND INEQUALITY

Climate change isn't a neutral force; rather, its impacts are distributed unevenly across populations, exacerbating existing inequalities and disproportionately affecting marginalized people and communities. While it's a disconcerting reality, as climate change accelerates, marginalized people will continue to endure the brunt of the societal and economic consequences stemming from climate change fallout.

How Climate Change Disproportionately Affects Marginalized Groups

Low-income populations, communities of color, ethnic minorities, indigenous groups, elderly people, young people, and other vulnerable demographics experience climate change in ways that magnify existing social inequalities. Four distinct disparities highlight the unequal burden these groups face:

- **Economic:** Socioeconomic status greatly determines a community's resilience to climate impacts. Marginalized communities often lack the resources to adapt to changes such as extreme weather events, sea-level rise, and reduced agricultural productivity. Economic vulnerabilities intersect with climate vulnerabilities, creating a cycle of disadvantage.
- **Racial:** Racial minorities are disproportionately affected by environmental hazards and pollution, leading to a phenomenon known as environmental racism. These communities often reside near hazardous sites or areas prone to environmental degradation, worsening their exposure to climate-related risks.
- **Global and Regional:** Low-GDP nations, often home to resource-poor, marginalized populations, contribute less to greenhouse gas emissions but suffer the most from climate impacts. These countries lack the resources to adapt effectively and are more susceptible to natural disasters and resource scarcity.
- **Generational:** Future generations will inherit the consequences of climate change, and marginalized

communities who were born after the climate tipping point will continue to be particularly vulnerable due to the compounding effects of ongoing injustices happening in the present day. Elderly populations in some countries are neglected by their families and failing state and local social support systems, leaving them more vulnerable to the effects of climate change than other populations.

Environmental Justice: What Does It Mean?

Environmental justice is a concept that addresses the unequal distribution of environmental benefits and burdens based on race, income, and other sociodemographic factors. This concept acknowledges that marginalized communities are more likely to experience negative consequences in the face of environmental hazards, pollution, and climate change due to deeply ingrained preexisting systemic inequalities.

The growing problems of global temperature rise, natural disasters, and changing weather patterns can compound and exacerbate pre-existing vulnerabilities. This compounded effect threatens to cripple communities lacking the economic means, material resources, or social capital needed to cope effectively in the face of disaster.

Addressing environmental racism and inequality requires climate policies that go beyond solely mitigating greenhouse gas emissions. Inclusive strategies that uplift marginalized communities must be created to ensure that they have access to clean air, clean water, and equitable opportunities for adaptation and resilience-building.

Climate solutions should not perpetuate or exacerbate existing injustices. Rather, they need to redress historical disparities, promote inclusive decision-making, and prioritize the well-being of marginalized groups. By integrating social equity into climate policies, we can work toward a more just and sustainable future where climate resilience, dignity, and empowerment are accessible to all, regardless of their socioeconomic background or identity.

CLIMATE REFUGEES AND DISPLACEMENT

The intensifying impacts of climate change have given rise to a distressing phenomenon: Climate refugees and internally displaced persons (IDPs). The challenges faced by human populations forced to migrate due to the adverse effects of climate change will only increase over the next few decades. By being sensitive to the distinct nature of this type of displacement compared to other forms of migration, we can be better prepared to address the political and social concerns raised by climate refugeeism while learning how to best prepare for the shifting populations and changing demographics that'll accompany future waves of climate-related mass migration.

Refugees and IDPs Vs. Migrants: Making the Distinction

Climate refugees and IDPs are people driven from their homes due to the direct or indirect consequences of climate change. Unlike traditional migrant groups who may seek better economic prospects or improved livelihoods, these groups are forced to move by the harsh realities of changing environmental conditions. Pushed from their homes by factors such as rising sea levels, extreme hurricanes or flooding, and ecological degradation, they are often already vulnerable populations pushed over the brink by the changing climate.

Labeling populations solely as "climate displaced" or "climate migrants" fails to fully encapsulate the severity of the dilemmas they face. The term "displaced" implies the potential for eventual return, which often remains unattainable for those affected by the impacts of climate change (The Third Pole, n.d.).

Being an immigrant myself—having been whisked away from my home country at the age of 12 by my mother—I feel especially sensitive to the plight of migrants and refugees of all types. Though immigrating to America had its own challenges for my family, IDPs and climate refugees encounter a myriad of challenges that other immigrant groups don't have to face. Displacement disrupts not only their physical lives but also their social, cultural, and economic ties to their home communities.

Access to essential resources such as clean water, food, shelter, and healthcare can become precarious for displaced people, exposing them to increased vulnerability. The places where they seek refuge may lack the infrastructure and resources to accommodate sudden population influxes, intensifying competition for limited resources and potentially leading to political conflict.

As the climate crisis deepens, the issue of climate-induced displacement becomes increasingly urgent. It's estimated that by 2050, there could be up to 1.2 billion climate refugees worldwide (McAllister, 2022). To address this looming migration crisis, crafting effective, compassionate responses to the plight of climate refugees and IDPs, ensuring their safety, dignity, and the restoration of their lives amidst the challenges of an altered world is crucial.

PURSUING CLIMATE JUSTICE ON ALL FRONTS

Climate justice can help illuminate the pathway to climate solutions that promote equity, fairness, and collective responsibility. There are a number of core principles that underpin the idea, emphasizing its significance in the fight against climate change and its far-reaching implications for our global community. In the following section, we'll lay out these principles in clear terms.

The Principles of Climate Justice

Climate justice isn't just a lofty concept. It's a set of principles that can guide us toward meaningful action and positive transformation. The principles that form the ethical backbone of climate justice include:

- **Respecting and protecting human rights:** At the heart of climate justice is the recognition and safeguarding of human rights. It's about ensuring that everyone's dignity, well-being, and safety are prioritized, especially for those most vulnerable to the impacts of climate change.

- **Supporting the right to development:** Climate justice envisions a world where every nation has the chance to pursue sustainable development. It acknowledges that addressing climate change shouldn't hinder economic progress in developing countries. Instead, it aims to foster fair and sustainable global growth.
- **Sharing benefits and responsibilities fairly:** Fairness and shared responsibility are key in the fight against climate change. Climate justice demands that the benefits and responsibilities of addressing climate issues be distributed fairly, regardless of national borders or economic differences. Wealthy, carbon-producing nations should be held responsible as the primary offenders.
- **Inclusive decision-making:** Making decisions about climate policies and actions should be an inclusive process involving a diverse range of voices, stakeholders, and communities. Transparency and accountability in these processes ensure that climate commitments are honored.
- **Gender equality and empowerment:** Climate justice acknowledges the specific challenges faced by women and other marginalized gender groups. It calls for policies that empower and amplify all voices, recognizing that inclusivity is the way forward.
- **Transformation through education:** Education can be a powerful catalyst for change, empowering ordinary people with knowledge and fostering a sense of

responsibility towards the environment. Climate justice emphasizes the importance of access to education and fair, unbiased information that can help us make more informed decisions and nurture a shared commitment to the planet amongst people from varied backgrounds.

I would love to hear from you!

If you found value in this book, I kindly ask for your
review and for you to share your experience with others. Your
feedback helps us spread the message of sustainability to even
more individuals.
It's through your support and reviews that my book is able to reach
the hands of other readers.

Please take 60 seconds to kindly leave a review on Amazon. Please
scan the QR code below. If you reside in a country that is not listed,
please use the link provided in your order.

**All it takes is 60 seconds
to make a difference!**

4

FORGING A SUSTAINABLE PATH
FORWARD

Climate change represents a challenge that extends beyond mere environmental concerns. It also carries with it profound economic implications that are going to become looming factors in the overall health of our global economy. The projections are alarming, painting a future where the unchecked effects of climate change could cost the world trillions of dollars by the close of the century (Deloitte, 2022).

The time to start envisioning a world where sustainable practices and economic growth are not adversaries but allies is now. Our ability to progress and survive as a society depends largely on our collective ability to grasp the gravity of climate realities while acknowledging the opportunities they present for new technologies, new jobs, enhanced global healthcare, and improved crisis response management. Profit-seeking ventures and environmental stewardship must come together. A genuine fusion between economic progress and real climate solutions could be the turning point that defines the world to be inherited by generations to come.

In the previous chapters, we've explored the relationships between climate change factors, biodiversity, justice, and ecosystems. We've spoken of the societal and social implications for human populations. Now, it's time to put on our accountant's visor and start tallying up the financial toll of climate inaction and show how solving the problems of tomorrow can bring us high profits while also saving the planet.

The uncharted potential of a green economy is huge. The choices we make within this nexus will reverberate far beyond balance sheets, shaping our well-being, supporting the vibrancy of communities, and bringing prosperity to underdeveloped nations in the decades to come. The interplay between capitalism and climate is complex, to say the least. It's a dynamic that begs us to weigh the consequences of inaction against the pitfalls associated with capitalism's baked-in tendencies toward resource and sometimes human exploitation.

THE ECONOMIC COSTS OF CLIMATE CHANGE

As the effects of climate change intensify, a critical spotlight shines on the economic costs that result from its consequences. From extreme weather events that pummel communities to rising temperatures that threaten vital economic sectors, the worldwide economic ledger is undergoing a transformation that necessitates a closer look.

Extreme Weather Events: The Ever-Growing List of Climate-Related Expenditures

As we've already discussed at length, scorching heat waves and wildfires wither crops, burn forests, and threaten human populations. Rising sea levels, torrential downpours, and flooding can sweep whole villages away. Whether we'd like to admit it or not, extreme weather events are on the rise. Not only do they continue to wreak havoc on our cities, towns, and natural landscapes, but they've also emerged as potent agents of economic disruption.

As climate change intensifies, the economic toll of extreme weather events becomes increasingly pronounced, setting the tone for an extremely costly, uncertain future. The true extent of the economic repercussions becomes evident in the aftermath of extreme weather events, as insurance adjusters rush in to assess the immediate damages, but what's the cost of extreme weather over time?

A comprehensive analysis by The White House (2022) called *The Rising Costs of Extreme Weather Events* revealed that the United States alone has witnessed a staggering $1.5 trillion in economic damages between 1980 and 2020 due to these types of events. Unpredictable and relentless, these weather events hit hard across various economic sectors, leaving upheaval in their wake.

Agricultural Losses

The agricultural sector, often portrayed as the lifeblood of nations, bears the brunt of these climatic assaults. Agriculture, the very foundation of our sustenance, has been dealt severe blows as crop yields are diminished, livestock is threatened, and the stability of food supply chains is shaken. The consequences can have a big impact on regular people, their communities, and national economies, threatening to leave a trail of disrupted livelihoods and inflationary prices.

Damaged Infrastructure

From crumbling roads to eroded coastlines, the physical fabric of nations can easily become strained and start to crack. The economic investment required for repairs and restoration in the future could be staggering as more and more extreme weather events rip through our cities and towns. Infrastructure serves as the backbone of economic progress, and the impacts of climate-induced wear and tear undermine long-term growth prospects, especially in already climate-disadvantaged, resource-poor developing countries.

Emergency Management Response and Healthcare

As temperatures rise and weather patterns become increasingly erratic, public health systems across the world are stretched to the limit. Escalating incidences of heat-related illnesses and the spread of vector-borne diseases are poised to become not only humanitarian crises but also economic ones in the future.

Heat-strained healthcare systems translate to reduced productivity and greater expenditure as the workforce battles the health consequences of a changing climate. While in the United States, climate risk to the healthcare industry is just starting to emerge as a threat, in other parts of the world, this concerning phenomenon has been clearly demonstrated in recent years.

A 2021 study published in Lancet found that one-third of healthcare workers in Australia fled from underserved areas they had previously worked in due to the threat of extreme heat fueled by climate change (Pendrey et al., 2021).

Travel and Tourism

Tourism is a big part of many economies, and by no means immune to climate impacts. The allure of destinations once celebrated for their natural beauty can quickly fade as weather events transform paradises into disaster zones. Take what's happening right now in Maui and Greece as examples of this recent paradise-in-flames reality we commonly confront. The occurrence of these natural disasters and weather events can easily hamper the inflow of tourists while disrupting local economies.

Productivity

Extreme weather events can lead to lost productivity for all types of workers, as workdays are curtailed, supply chains are interrupted, and infrastructure and transportation systems that get people to their jobs falter under the strain of disaster conditions.

Workers forced to contend with road closures, power outages, or damaged public transportation systems face significant hurdles in commuting to their workplaces. Businesses, especially those in vulnerable sectors, could experience heavy disruptions in production and delivery schedules, further exacerbating supply chain woes caused by climate damage.

Business and Finance

The business and financial sectors stand vulnerable to the economic tempests of climate change. Heightened risks from extreme weather events can trigger a domino effect, impacting not just individual businesses, but also the broader financial markets. Uncertainty caused by an uptick in extreme climate change-fueled weather patterns breeds wary attitudes among investors and consumers alike, leading to cautious spending patterns and slower economic growth—potentially leading to market volatility or even increased frequency of stock market crashes.

The Societal Toll

When uncertainty clouds the horizon, human behavior, thoughts, and mental health can all be affected. The growing awareness of climate change's economic implications has already cast a shadow of doubt over people's belief in opportunities for the future, provoking a prevalent sense of climate anxiety. This uncertainty and the anxiety that often accompanies it has the potential to resonate throughout society, influencing the job market, consumer behaviors, corporate strategies, and policy decisions.

As families and businesses brace for the unknown, a collective wariness can set in—as dark, sweeping sentiments threaten to fundamentally alter economic landscapes. The societal toll of these wary attitudes encompasses not just financial matters but also broader aspects of health and well-being, trust in government and institutions, and long-term financial planning.

GREEN BUSINESSES, JOBS, AND SUSTAINABLE ECONOMIC GROWTH

Amidst the challenges posed by the climate crisis, a silver lining emerges—the potential for a transformative shift toward a more sustainable economy. Too often, the discourse surrounding climate action centers on the perceived sacrifices required on our part. However, it's not just about individual responsibility, it's about our need to rally together.

History has shown that crises can act as catalysts for profound societal changes and new opportunities. Think about the period of rebuilding and economic progress that came in the wake of World War II. Not only did it propel the United States and Western Europe to new economic heights, but the advanced communication tools brought to us by revolutions in information science and computing during this time turned out to be the key technologies of our age. The climate crisis is no exception: From crisis, progress, change, and big ideas that revolutionize the future can be born.

By embracing sustainable practices, today's businesses and national economies can help mitigate environmental harm while paving the way for innovation, job creation, and sustainable economic growth. Let's look at the ways that this challenging feat can be accomplished.

Business Opportunities

In the face of the climate crisis, new avenues for business and innovation are opening up. As industries transition to cleaner energy sources and environmentally friendly practices, innovative solutions are sought-after commodities. These solutions span from renewable energy technologies to waste reduction strategies and sustainable manufacturing processes. Businesses that proactively address environmental challenges can effectively position themselves for long-term success by fostering resilience, reducing operational costs, and tapping into growing markets for sustainable products and services (Enel Green Power, 2023).

Intergovernmental agencies such as the World Bank have emphasized that adapting to climate change can be both a threat and an opportunity for the private sector, emphasizing the pivotal role of private businesses in the transition to a low-carbon future (Tsitsiragos, 2016).

New Job Creation

One of the most significant benefits of the transition to a green economy is the creation of new jobs. The push toward renewable energy, energy-efficient technologies, and sustainable practices generates a higher demand for skilled workers in various sectors. For instance, electrification and renewable energy adoption lead to less pollution and cleaner air, resulting in improved public health and reduced healthcare costs.

Moreover, the shift towards green technologies fosters energy independence and savings by reducing reliance on fossil fuel imports. These aspects contribute to economic stability and resilience, creating a positive cycle of growth and prosperity.

Investment Opportunities

Unique investment opportunities will come about across various sectors if the private sector takes the problem of solving the climate crisis seriously. The transition to renewable energy sources for electricity generation will offer investors a chance to tap into a growing market with significant potential for returns. New, sustainable ways of investing are set to emerge.

Carbon credit and offset trading has already become a popular way for companies and individual investors to balance their carbon footprint through the markets. Based on the "cap and trade" regulatory approach that helped reduce sulfur emissions in the 1990s (Kenton, 2019), investment schemes have been shown to have a real impact on mitigating emissions levels.

Aside from finance mechanisms that can bring carbon offsets, industries like construction infrastructure are likely to do their part in carbon reduction by introducing new energy-efficient building technologies. Green energy companies will also hopefully start gaining a larger market share in the coming decades to further aid the transition.

While the prevalence of electric vehicles is rising, the use of high-quality refined fuels is not going to cede. This makes biofuels another promising area that could represent a great investment vehicle for the coming decades. As fossil energy gets further capped—hopefully becoming obsolete—more and more green fuel alternatives will be needed to replace coal and natural gas.

Tourism and air transportation are two other industries that could greatly benefit from introducing more sustainable practices, capturing an already strong consumer demand for eco-friendly travel options.

A Shifting Workforce: Green Job Opportunities for the Future

In transitioning to a green economy, the emergence of green jobs will play a crucial role. These jobs encompass a wide range of roles that contribute to sustainable practices, clean energy adoption, and environmental protection. Examples include renewable energy technicians, energy auditors, sustainability managers, and eco-friendly product and service designers.

Green jobs not only provide employment opportunities but help further drive innovation while contributing to economic growth and promoting the shift toward a more sustainable future. These new employment opportunities represent a cornerstone of the transition, aligning workforce needs with environmental imperatives.

Amidst the growing demand for green jobs in the transition to a sustainable economy, plenty of job seekers are already looking to pursue meaningful careers while contributing to environmental stewardship. There are a few ways that green job candidates can take strategic steps to secure positions in these emerging fields.

If you're someone who has a passion for sustainability and strategic planning skills, do the research and start networking!

If you have the drive and the right skill set, you, too, can position yourself to thrive in the dynamic world of green jobs. As the green economy continues to grow, landing a green career can help you contribute to your professional growth, increase personal satisfaction, and meet the broader goal of helping achieve a sustainable future for all.

Green Business Success Stories

Numerous big businesses have recognized the financial benefits of embracing environmental responsibility. This realization has led to innovative strategies that not only reduce ecological footprints but also bolster profitability.

United Airlines, for example, has invested in reducing the tare weight of its aircraft to lower emissions. By auditing all excess equipment, materials, and items (even down to seemingly insignificant items like sheets of printed paper) carried on board, they were able to trim down on weight, cutting down on fuel costs and emissions.

As many hotel chains have done in recent years, Clarion Hotel Group has mounted initiatives that encourage guests to cut down on energy usage by reusing towels and foregoing cleaning services for short stays. They also redesigned their bathroom fixtures to be more efficient.

Retail giant Walmart has committed to reducing emissions and promoting sustainability throughout its supply chain by fitting its fleet of trucks with software that reduces emissions and provides for more fuel-saving route planning.

Technology giants Google, Facebook, and Amazon have also made strides towards operating with lower environmental impact, pledging to make their data centers 100% reliant on renewable energy (Martin & Dent, 2019).

These successful corporate initiatives show that environmentally conscious practices can align with profitability, illustrating the potential for win-win scenarios in the business world.

As the climate crisis spurs transformative change, businesses have a unique opportunity to lead the charge toward a sustainable future. By embracing innovative solutions, creating green jobs, and reaping the economic benefits of environmentally responsible practices, the business world can help forge a path toward prosperity that doesn't compromise the health of the planet or the well-being of its inhabitants.

INVESTING IN CLIMATE SOLUTIONS

It's not just big businesses that have the power to use their financial resources to help make more climate-friendly choices for the future. Believe it or not, you, too, have the opportunity to leverage your personal financial choices for the betterment of the planet. Your personal investments, including your 401K, can be a great place to start looking for green investment opportunities.

Any eco-friendly investor should seriously consider divesting from fossil fuel-heavy holdings and reallocating those funds toward renewable energy companies or innovative green technology startups. There are a number of strategies and approaches available to individuals seeking to make environmentally conscious investment decisions; here, we'll cover a few of them.

Tools and Strategies for Investors

Investing in climate solutions requires a strategic approach that aligns your financial goals and investment timeline with the effects of rapidly accelerating climate change. When considering new investments, there are a number of different rankings you can look at.

One of the best ways to look at a company's environmental impact is by looking at their environmental, social, and governance (ESG) criteria. Financial services companies such as Morningstar provide sustainability ratings based on this ESG data, which can help make it easier to interpret the ESG data and assess the carbon impact of companies you're considering investing in.

These types of sustainability rankings help you make sure you're choosing to invest in companies that are truly committed to addressing climate challenges while promoting positive societal change. Another way that investors can go green is by considering investing in clean energy funds or exchange-traded funds (ETFs) that focus on renewable energy sources, low-carbon technologies, and sustainable practices.

The Challenges of Climate-Focused Investing

Making the conscious choice to engage in climate-focused investing is a noble aspiration that can help contribute to a healthier planet and a more sustainable future. Yet, when it comes time to start planning a green investment portfolio, investors are often confronted with a fundamental challenge: the delicate equilibrium between financial returns and the pursuit of impactful change.

One of the primary challenges lies in the perception that environmentally conscious investments might compromise financial gains. Skepticism often stems from concerns that companies committed to environmental sustainability may divert resources away from profit-generating activities, potentially leading to subpar returns. In this context, understanding that sustainable practices and profitability can coexist becomes imperative.

Furthermore, the relative newness of climate-focused investing can introduce uncertainties. Traditional financial analysis tools might not be fully equipped to evaluate the impact of environmental factors on an investment's performance. This information gap can create ambiguity for investors seeking to make informed decisions.

Aside from looking at the ESG data of individual companies that make up your portfolio, aligning with socially responsible investment (SRI) strategies can provide a framework for evaluating investment options. SRI considers not only financial returns but also the broader societal impact of investments. This approach allows investors to proactively shape their portfolios in ways that align with their values.

While the challenges of climate-focused investing are undeniable, there are no significant barriers preventing actionable strategies that can lead to great returns. The quest for profitable and impactful investments is feasible through the adoption of prudent strategies, thorough research, and an understanding that a commitment to environmental sustainability need not come at the expense of financial gains.

How to Tackle Climate Change in Your Investment Portfolio

As we already discussed, reallocating funds in your investment portfolio away from fossil fuels and into sustainable ventures is one of the most direct ways to combat climate change through your asset allocation. By leaving companies either involved in fossil fuel extraction or that otherwise have massive carbon footprints, you'll not just feel good about your choice but will also be helping promote the transition to a low-carbon economy.

What Is Impact Investing?

Impact investing introduces a dynamic approach where your investments are directed toward projects that generate financial returns and address environmental and social challenges. This strategy emphasizes proactive engagement with companies to foster sustainable practices, shaping corporate behavior through investment decisions.

Types and Examples of Impact Investments

Impact investing enables you to support causes you're passionate about while aiming for returns that match your financial goals and investment timeline. Here are a few examples:

- **Renewable energy ventures:** These investments channel capital into renewable energy projects such as solar, wind, and hydroelectric power. The financial returns are often linked to the production and sale of energy, aligning profits with a cleaner energy future. Consider investing in a solar energy company as a great starting point.

- **Sustainable real estate:** Impact investments in this category target properties and developments with strong environmentally conscious features. This might include energy-efficient buildings, sustainable materials, and green spaces. Consider buying shares, for example, in a real estate fund that focuses on developing LEED-certified commercial buildings.

- **Affordable housing initiatives:** These investments contribute to providing affordable housing solutions for underserved communities. Such projects can improve living conditions and alleviate housing disparities. You might consider investing in a fund that finances the construction of affordable housing units for climate refugees and IDPs as a great way to put your money where your heart is.

- **Clean technology startups:** This category focuses on early-stage companies that develop innovative technologies to address environmental challenges, from waste reduction to sustainable agriculture. Consider investing in a startup that works on developing new technologies that convert organic waste into renewable energy, for example.

When you consider making impact investments, you're really putting your money where your mouth is, aligning your financial goals with your ethical, environmental, and social values. The diverse range of sectors where it's possible to engage in these types of strategies showcases the potential for diversification across a range of hand-chosen impact investments. As the capital you deploy helps drive real, meaningful change, you'll be glad you made the choice to invest in addressing climate change.

Climate Change Action Investment Approaches

Investment bank Morgan Stanley has defined three guiding principles for investment approaches that help combat climate change. Here they are:

- **Intentionality:** Minimize exposure to companies linked to greenhouse gas energy sources and activities, such as coal, oil, and gas. Include ESG data in your investment screening process. Consider investing in decarbonization technology.
- **Influence:** As a shareholder in any company, you have the right to influence the decision-making processes and be an advocate for positive, sustainable change. Don't forget it!
- **Inclusion:** When investors recognize the interconnectedness of climate change, racial justice, and gender equality, they can use their investments to promote diversity and inclusive practices (Morgan Stanley Wealth Management, 2022).

To Sum Things Up

Green businesses, jobs, and sustainable economic growth have emerged as key avenues for creating a prosperous and environmentally conscious future. The increased interest in climate-friendly investment products is a step in the right direction, but at the same time, it doesn't represent a viable long-term solution to the root problems at hand. While confronting these issues through our 401ks can be a meaningful gesture, the work toward climate justice doesn't stop with allocating your financial resources in more climate-friendly ways.

The next crucial step to take in tackling climate change is minimizing our own carbon footprints by reducing our resource consumption. This is where The Eco Ripple System comes in, helping provide guidance to us while making greener lifestyle choices that can collectively make a significant impact. In the next chapter, we'll be looking at practical strategies to incorporate this system into your life, helping you adopt eco-friendly habits and practices that can benefit the planet while leaving you feeling personally fulfilled.

5

FROM RIPPLES TO WAVES

E very day, from the moment you wake up to the choices you make throughout your day, you have the power to shape the world around you. It's easy to underestimate the impact of our daily actions, but the truth is that every step we take leaves an imprint on the planet.

In a world driven by consumerism and convenience, it's difficult to keep green objectives at the forefront of our minds. However, making a few minor changes in our choices and habits can have a remarkable ripple effect, leading to transformations that extend far beyond ourselves. The question that arises is this: Will you choose to be part of the solution, or will you go on inadvertently contributing to the problems our planet faces?

MINIMIZING YOUR CARBON FOOTPRINT AND THE BENEFITS OF REDUCING RESOURCE CONSUMPTION

As we go about our daily lives, shopping for groceries, commuting to work, and enjoying our free time, we can use all these quotidian activities as opportunities to help the environment. By making mindful decisions, we can not only offset our individual impact on the planet but can also contribute to larger, positive changes.

THE ECO RIPPLE SYSTEM

The Eco Ripple System presents a promising and empowering framework for fostering a more sustainable future. This innovative approach to consumption and climate impact revolves around the idea that the journey toward sustainability begins with the choices we make on an individual level.

By embracing the idea that every action, no matter how small, contributes to a larger climate reality, we can become catalysts for positive change within our environment and the greater society we're part of.

At the heart of The Eco Ripple System lies the notion of personal responsibility. It underscores the idea that each of us possesses the power to create significant ripples of positive change. While it's easy to become overwhelmed by the enormity of global environmental issues, this system shifts the focus back to the individual, not in a blame-based way, but to highlight the fact that the journey towards sustainability starts within each of us.

The profound impact of seemingly small choices is the central theme of this system. It reminds us that even modest actions can lead to transformative outcomes when multiplied across communities and nations. By embracing conscious decision-making and sustainable practices, we can contribute to a collective movement of positive change, wherein our actions resonate far beyond ourselves and our own tiny corners of the world.

The Eco Ripple System can serve as a guiding doctrine, urging us to be actively aware of the choices we make in our daily lives. Whether it's opting for reusable products, reducing energy consumption, or supporting environmentally friendly practices, each decision has the potential to amplify the ripple effect.

Throughout the following sections, we'll lay out the key principles of this system, offering insights and practical guidance on how to integrate them into your life. Remember: When we embrace The Eco Ripple System, we set forth a chain reaction that reverberates throughout our surroundings, positively affecting others and the world around us.

Reducing Energy Consumption at Home

Making your home more energy efficient offers climate-conscious homeowners a win-win scenario as reducing energy consumption translates to both cost savings and decreased carbon emissions. Selecting energy-efficient appliances is the first crucial step.

Appliances

When it comes to making the right selection, considering factors such as labeling, size, and functionality can make a substantial difference. Opting for appliances with the Energy Star label or similar certifications ensures that the products adhere to high energy efficiency standards. Furthermore, choosing appropriately sized appliances and utilizing them in eco-friendly ways can lead to substantial energy savings over time.

With larger kitchen appliances, like dishwashers and refrigerators, choosing models equipped with energy-efficient features can make a big difference. From dishwasher models that minimize water and energy consumption to refrigerators designed with smart, eco-friendly technology, our selections can impact both our utility bills and the environment.

Choosing and Using Energy-Efficient Appliances

Selecting and utilizing energy-efficient appliances can make a big difference in minimizing our carbon footprint and reducing energy consumption. By adopting mindful practices and making informed choices, we can actively contribute to a greener, more sustainable future. Here are a few more essential steps to guide our decisions on and usage of home appliances:

- **Look at labels:** As already mentioned, labels such as Energy Star provide valuable information about the energy efficiency of appliances. These labels help us identify products that meet rigorous energy performance standards and can substantially reduce our energy consumption over time.

- **Choose the right-sized appliances for your home and household size:** Selecting appliances that match our needs and usage patterns ensures optimal efficiency. Oversized appliances not only consume more energy but can also lead to unnecessary waste of resources.
- **Find eco-friendly ways to use appliances:** Beyond the choice of appliance, our usage habits significantly impact energy consumption. Simple practices contribute to energy savings, such as using the dishwasher only for full loads, washing clothes in cold water, and minimizing dryer usage.
- **Consider smart appliances:** Smart technology allows for remote monitoring and controlling of energy usage through our mobile devices. Smart appliances can dynamically adjust settings according to our needs, optimizing energy efficiency.
- **Choose an energy-efficient dishwasher:** When selecting a dishwasher, opt for models with features that enable water and energy conservation. Efficient dishwashers often have shorter wash cycles and sensors that adjust water usage based on load size.
- **Pick a refrigerator with eco-features:** Refrigerators are among the most energy-consuming appliances in our homes. Opt for models with eco-friendly features such as adjustable temperature settings, high insulation levels, and automatic defrost.

- **Make cooking energy-efficient:** Utilize energy-efficient cooking methods such as induction ranges, pressure cookers, slow cookers, and toaster ovens.
- **Use an electric kettle for boiling tap water:** Electric kettles are an efficient way to boil water for beverages or cooking, as they heat water more quickly than conventional stovetops.
- **Make laundry day more energy-efficient:** Embrace practices that optimize energy usage during laundry. Washing clothes in cold water, using a clothesline or drying rack instead of a dryer, and ensuring that the washing machine is full before starting a cycle all contribute to energy savings.

Simple practices mentioned above, like using a kettle for boiling tap water, might seem insignificant, but they really can significantly decrease energy consumption over time! Each choice we make in favor of energy efficiency contributes to a positive ripple effect, fostering a more harmonious relationship between our lifestyle and the environment.

Practical Tips for Saving on Home Energy Costs

Beyond appliance choices and our habits when using them, there are a bunch of other ways you can implement energy-efficient practices in your own home. From sealing gaps and cracks in windows and doors to using programmable thermostats, steps like these collectively create a substantial impact on our home heating bill. Installing LED lighting, reducing vampire energy by unplugging appliances when not in use, and regular maintenance of heating and cooling systems are a few other great ways to show your commitment to energy efficiency while saving money over time.

By embracing these energy-saving strategies within our homes, we not only reduce our environmental impact but also become instrumental in advancing the goals of sustainability. Every choice made in the direction of energy efficiency serves as a testament to our commitment to creating positive ripples of change within our environment and greater society.

Adopting Eco-Friendly Forms of Transportation

Transportation choices have a big influence on our carbon footprint and the health of our planet. By embracing sustainable transportation options, we can drive positive change while contributing to a greener future. In this section, we'll take a look at the environmental consequences of our travel choices and highlight some of the more eco-friendly modes of travel.

The Benefits

Sustainable transportation options offer a range of environmental benefits that ripple far beyond our individual choices. By transitioning to eco-friendly modes of travel, we can collectively contribute to:

- **Less pollution and clearer skies:** Vehicles, particularly those powered by fossil fuels, emit pollutants that degrade air quality and contribute to climate change. Opting for sustainable transportation helps reduce these emissions, resulting in cleaner air and improved public health.

- **Healthier communities:** Decreased emissions from sustainable transportation translates to improved air quality. Lower air pollution levels lead to healthier communities as there are lower incidences of chronic respiratory ailments like asthma (Jiang et al., 2016).
- **Reduced levels of harmful chemicals:** Sustainable transportation options often involve cleaner fuels and technologies that emit fewer harmful chemicals, benefiting both the environment and human health.
- **Fewer cars equal more green spaces:** Embracing alternatives to personal car usage reduces the need for expansive road networks, parking lots, and other paved surfaces, preserving more natural landscapes.
- **Noise pollution reduction:** Sustainable modes of travel, such as electric vehicles and public transit, generate less noise pollution, contributing to quieter and more peaceful urban environments.

What Are the Most Eco-Friendly Ways to Travel for Commutes and Short-Distance Travel?

- **Biking:** Cycling is an environmentally friendly option that promotes physical activity and reduces traffic congestion. Investing in a sturdy bike and using cycling lanes or dedicated paths can make your commute safe, eco-friendly, and enjoyable. Don't forget to wear a helmet too!
- **Electric bikes or scooters:** Electric bikes combine the benefits of cycling with a boost from an electric motor. They offer an energy-efficient mode of travel,

particularly for hilly terrains or longer distances. These days, e-bike and electric scooter rentals are all over many cities, and all you need is an app. Consider giving them a shot on your commute or on your lunch break.

- **Trains and subways:** Utilizing public transportation systems such as commuter rail lines, subways, buses, trams, and trolleybuses are all effective ways to reduce individual carbon emissions while contributing to more efficient use of urban space. Bus fleets in some cities, such as San Francisco, are also undergoing modernization, with a substantial part of their fleet being replaced by new electric vehicles.

- **Carpooling:** Sharing rides with others headed in the same direction as you reduces the number of vehicles on the road and helps cut down individual emissions.

What Are the Most Eco-Friendly Ways to Travel Long-Distance?

- **Bus:** Buses are one of the most energy-efficient ways to travel longer distances. While electric buses are best, modern hybrid or conventional combustion engine buses often have lower emissions and provide a budget-friendly means of transportation.

- **Train:** Although train infrastructure in the United States is lagging compared to many other countries, Amtrak trains are highly energy-efficient compared to commercial aircraft and offer a scenic way to cover longer distances. They also are much better on emissions when compared to personal car travel or short flights.

- **Electric Vehicles:** Electric vehicles (EVs) are an increasingly popular choice for environmentally conscious travelers. With zero tailpipe emissions, EVs contribute significantly to reducing carbon footprint. As the proud owner of an EV myself, I can't say enough about how much it's made a difference in my life.
- **Ridesharing:** Car-sharing services and rideshares offer a convenient and environmentally friendly option for long-distance travel. Allowing for less car ownership overall and multiple passengers to share rides together.

What Are the Most Eco-Friendly Ways to Travel Internationally?

When traveling overseas, flying is often the most common choice, but it has a substantial environmental impact due to high levels of emissions. While it's possible to book overseas passage on cargo ships (yes, really—I've even considered doing it myself before), this can be quite expensive and can turn an hours-long trip into multiple weeks.

If you're on a multi-stop international trip, once you've landed at your first location, try to avoid any further air travel whenever possible—except for your return flight, of course. Instead, choose more sustainable options like train or bus travel whenever possible to get to your next stops.

To Sum Things Up

By making informed travel decisions and embracing sustainable transportation options, we can individually and collectively contribute to a more sustainable future. Every eco-friendly choice we make sends positive ripples through our environment, creating a collective impact that fosters change and protects the planet.

CONSERVING WATER

Water is our most vital natural resource, and its conservation is a crucial part of our responsibility toward the planet. Water is considered a renewable resource, as groundwater is naturally purified as it passes through underground rock and mineral deposits. However, this natural process is ever-threatened by fossil fuel mining processes such as fracking, chemical leakage, and other sources of groundwater contamination. Saving water matters, and it's not just to keep our utility bills down.

Why Should We Care About Saving Water?

Water scarcity is becoming an increasingly pressing issue in some places, particularly those already affected by drought. Regions come to face water resource stress due to over-extraction, pollution, and climate change. Water conservation helps alleviate this stress and ensures that future generations will continue to have access to clean, natural drinking water sources.

Saving water reduces energy consumption associated with water treatment and transportation, cutting greenhouse gas emissions and reducing the carbon footprint of water utility companies. By using water more efficiently in our homes and businesses, we can protect ecosystems, preserve biodiversity, and help secure the well-being of communities worldwide.

How Much Water On Average Do American Families Waste?

In the United States, water might seem abundant, but wasteful household habits contribute to substantial loss. On average, American households can save around 20%-30% of their total water usage by addressing leaks, inefficient appliances, and other careless behaviors (U.S. EPA, 2017). Inefficient use and waste of water resources not only strain local water systems but also create more carbon-producing energy demand and drain financial resources. By eliminating these common sources of water waste, we can significantly reduce our collective water footprint.

To put things into context, most Americans have never had to face water shortages and feel their real impact. I first became aware of shortages being a real issue in my 20s. When traveling through Venezuela, I stumbled across a long line of people, including kids, with various containers waiting for what turned out to be drinking water. I remember being shocked, especially given that I distinctly remembered passing what appeared to be a golf course being watered with a sprinkler system earlier the same day. Just like many other precious resources across the globe, water rights are often diverted to those at the top of society, leaving people on the lower tiers of the socioeconomic ladder to suffer.

What's the Main Purpose of Water Conservation?

The primary goal of water conservation is to ensure a sustainable supply of clean water for current and future generations. Beyond individual consumption, water conservation efforts also play an important role in maintaining freshwater and brackish ecosystems, supporting agriculture, and preserving natural habitats.

By using water efficiently and minimizing waste, we contribute to the health of our environment and the well-being of communities around the world. Water conservation efforts are based on the idea that our actions have a direct impact on the delicate balance of water resources. Today, it's more important than ever that we take responsibility for our role in safeguarding it as a precious, natural asset.

MINIMIZING HOUSEHOLD WASTE AND RECYCLING

A big part of The Eco Ripple lifestyle is making an effort to cut down on household waste and to be more vigilant about household recycling. There are a number of tips that can help you reduce everyday waste, and we'll be going through some of them in this section.

The Importance of Waste Reduction

Waste reduction is an important part of being a good citizen. By producing less waste, we can help decrease the strain on landfills, which have historically been major contributors to pollution and environmental degradation. By focusing on overall waste reduction and brushing up on our recycling skills, we can help support healthier and safer living conditions for all members of our communities.

The Benefits of Waste Reduction

- **Reduces landfills:** Waste that ends up in landfills can contaminate soil and water, posing serious health risks to nearby communities. By reducing waste, we alleviate the burden on landfills and reduce the potential for pollution.
- **Saves energy and resources:** The production, transportation, and disposal of goods consume significant energy and resources. By reducing waste, we reduce the demand for raw materials, energy, and water, leading to a more efficient and sustainable use of resources.
- **Makes for a safer future:** Proper waste management minimizes the release of hazardous substances into the environment, safeguarding the well-being of both humans and ecosystems. It contributes to creating a cleaner and safer environment for future generations.
- **Helps mitigate climate change:** Landfills are a significant source of greenhouse gas emissions, including methane, a potent contributor to global warming. By reducing the amount of waste we produce, we can help curb methane emissions.

Practical Ways to Reduce Waste

- **Use reusable shopping bags:** Single-use plastic bags contribute to plastic pollution and harm marine life. Switching to reusable shopping bags reduces plastic waste and encourages sustainable consumer behavior. Even the terrorist group Al-Qaeda has recognized that disposable plastic bags pose a real threat, banning them from use in 2018 for posing "a serious threat to the well-being of humans and animals alike" (Khalid, 2023).

So, if the guys behind 9-11 are expressing a sense of urgency over the dangers of plastic, you have to scratch your head and wonder: Does this make those still armed with a disposable plastic bag "climate terrorists?"

- **Use a reusable bottle:** Your reusable water bottle can serve as a constant companion. Bringing your beverages from home helps you save money and reduce waste. By opting for a reusable bottle or cup, you avoid purchasing drinks in single-use containers, which require a significant amount of energy for production and transportation.
- **Reduce food waste:** Plan meals, store food properly, and use leftovers creatively to reduce food waste. Composting food scraps also helps by diverting methane-producing organic matter from landfills.
- **Embrace near-zero waste consumption habits:** Aim to minimize the waste you generate further by being mindful of your consumption habits. Purchase products

with minimal packaging, choose bulk options that aren't individually wrapped, and try to buy only items that can be easily recycled or composted.

- **Shop local farmers markets:** Opt for local farmers markets to support local producers and reduce the carbon footprint associated with long-distance transportation. Buying in bulk reduces packaging waste and helps you save money in the long run.
- **Compost your organic household waste:** Composting not only reduces waste but also provides you with nutrient-rich soil for gardening. Starting your own compost heap at home is a simple and effective way to contribute to a circular economy.
- **Avoid single-use items:** Opt for reusable alternatives to single-use items like water bottles, coffee cups, and cutlery. These small changes significantly decrease the amount of waste produced.
- **Use less paper:** Reduce paper waste by going digital whenever possible. Opt for online subscriptions, digital receipts, and e-statements to minimize your reliance on paper. When printing is necessary, use both sides of the paper and recycle responsibly.
- **Recycle correctly:** Understand your local recycling guidelines to ensure that you're recycling effectively. Improper recycling can lead to contamination and hinder sorting and processing operations.
- **Repair and reuse:** Before discarding items, explore repair options to extend their lifespan. Consider repurposing or donating useful items that you no longer need.

- **Buy second-hand items:** Thrift stores, online marketplaces, and garage sales offer a wealth of options without the carbon toll associated with buying and shipping new products. Don't forget to donate clothing and household items you no longer need to extend their lifespan and prevent unnecessary waste.

By adopting these waste reduction strategies, you'll not only contribute to a healthier planet but can also inspire others to join the movement towards a more sustainable future.

Does Recycling Really Help?

Recycling is absolutely beneficial for the environment as it conserves natural resources, reduces the need for raw manufacturing and industrial materials, and decreases energy consumption. By cutting down on the amount of waste that makes its way to landfills and incinerators, recycling helps mitigate the negative environmental impacts of the overall waste disposal industry.

Moreover, recycling saves energy because it requires less energy to process some recycled materials than it does to produce new materials. For instance, recycling aluminum cans uses 95% less energy than producing cans from raw materials (ASM Metal Recycling, 2022). This energy savings translates to reduced greenhouse gas emissions and helps combat climate change.

Debunking Common Recycling Myths

Waste Management Inc. has published a PDF helping clear up common myths and misconceptions around recycling. Here are a few of the points they make that are worth considering:

- **Most Americans already recycle all they can:** While many Americans actively recycle, there's room for improvement. Having recycling bins in multiple rooms of the house, not just in the kitchen, can help make sure recyclable objects aren't sipping through the cracks.
- **Recycling arrows on packaging mean it's recyclable:** The recycling symbol (Mobius loop) doesn't necessarily indicate that an item is recyclable in your community. Recycling programs vary by location, so always check your local guidelines.
- **Containers must be cleaned out thoroughly:** While it's good to rinse food residue out of containers, they don't need to be spotless. Emptying and rinsing containers help prevent contamination, but thanks to improved processing at recycling centers, these days, they don't need to be sparkling clean.
- **Haulers will sort out items that don't belong:** Recyclables need to be correctly sorted to ensure efficient recycling. While recycling facilities do have sorting technology, proper sorting at home reduces the risk of contamination.
- **All plastic items are recyclable:** Not all plastic items are recyclable. Recycling facilities usually don't accept items like hoses, tubes, and shower curtains. If you're

not sure about an item, make sure to check your local recycling guidelines.

- **Aerosol cans can be recycled:** While most recycling facilities do accept pressurized aerosol aluminum, steel, and mixed metal cans, they must be empty and dry, with the plastic caps removed.
- **All glass bottles and jars are recyclable:** While glass bottles and jars are generally recyclable, it's important to follow local guidelines. Some facilities may not accept certain types, such as window glass (Waste Management, n.d.-a).

How to Properly Recycle All Kinds of Materials

- **Paper and cardboard:** Flatten cardboard boxes and watch out for any contaminants like pizza grease stains and stray cheese strands.
- **Plastics:** Check local guidelines to determine which types of plastic are accepted. While rinsing and removing the caps before recycling plastic bottles was the common practice for years, new guidelines from the Association of Plastic Recyclers say that the current recommendation is to rinse the bottle and then replace the cap.
- **Glass:** Clean glass bottles and jars are usually recyclable. Metal lids from mason jars and jam jars are also recyclable, so make sure to remove them and set them aside before recycling your glass containers.
- **Aluminum:** Empty, rinsed cans are always accepted. For aerosol cans or other containers that may have

traces of hazardous materials, it's always best to check with local authorities. You can also consider redeeming your cans if you go through a lot of them. All those nickels and dimes could add up to more than pocket change, and you could always donate it to a climate-friendly charity to offset the impact of your canned beverage habits.

- **Batteries:** Many recycling centers accept standard-type cell batteries, but most larger batteries, such as car batteries, portable power blocks, laptop batteries, and larger batteries, such as those for electric wheelchairs, require special handling for proper recycling and disposal.

- **Electronics:** Most home appliances, portable electronics, and technology items contain hazardous materials and should not be thrown out or recycled without following strict guidelines. Look for local electronic waste collection sites to recycle these items properly.

- **Food and yard waste:** While recycling facilities don't generally accept household food waste, leaves, or lawn trimmings, some recycling providers in certain municipalities may offer curbside compost and organic materials pickup.

- **Used oil:** Used motor oil and transmission fluid can be recycled at designated collection centers. Most auto parts stores are required by law to have collection barrels—just ask, and they'll often allow you to safely drop off your dirty fluids for recycling.

- **Household hazardous waste:** Dispose of household hazardous waste like solvents and chemicals at designated waste facilities. If you're not sure how to find one of these facilities or correctly dispose of chemical liquids or other potentially toxic materials, contact your local municipality.

- **Tires:** Most tire and auto part retailers accept used tires for recycling. Shredded recycled tires can be used for a number of secondary purposes, one of the most common being artificial turf soccer fields.

- **Scrap Metal:** Clean, uncontaminated scrap metal items can be recycled, but your curbside recycling program may not accept scrap metal in their bins. It's best to contact them if you're in doubt, and if you're dealing with any sharp or rusty metal pieces, make sure to exercise caution.

- **Miscellaneous:** For larger or mixed material items like garden and patio furniture, contact your local recycling services provider to see if there's any way to recycle these types of items. Items like above-ground pools and hot tubs will often require professional removal and dismantling before any parts can be salvaged or recycled.

MAKING SUSTAINABLE FOOD CHOICES

The things we eat can be a sensitive subject for some, from picky eaters to people who simply like what they like and aren't open to changing their dietary habits. Still, a lot of us don't eat a diverse range of locally available seasonal foods as our ancestors did. Instead, we rely on a variety of often imported produce flown in from all around the world. You can just imagine the climate impact this has!

The Oft-Neglected Environmental Impact of Our Food Choices

While the environmental impacts of industries like utilities, waste disposal, and transportation are frequently discussed, the contribution of our personal food choices too often goes unnoticed. The agricultural processes and food manufacturing industries that produce the foods we enjoy have significant carbon footprints, impacting land, water, and energy resources. Adopting sustainable food shopping and consumption habits is not only beneficial for your health but also for the planet's well-being.

What Types of Food Contribute The Most to Environmental Issues?

Certain types of food have a larger environmental impact due to their resource-intensive production methods. As we've already discussed, foods like meat and dairy, particularly from conventional farming, contribute significantly to climate change through methane emissions.

Farming methods of all types require extensive land use and water consumption and contribute to greenhouse gas emissions. By understanding which foods have a higher environmental toll, we can make more informed choices and attempt to mitigate our environmental impact.

Ten Simple Things You Can Do to Lessen the Environmental Toll of Your Diet

- **Know what you're eating and where it came from:** Awareness is the first step. Educate yourself about the origins of your food, production methods, and associated environmental impacts. Labels such as "organic," "non-GMO," and "fair trade" can provide valuable insights into the production and sourcing of the foods we buy.
- **Support sustainable agriculture:** Opt for foods produced through sustainable farming practices. Look for certifications like USDA Organic or Rainforest Alliance that emphasize environmentally friendly cultivation methods.
- **Avoid unnecessary packaging:** Excessive packaging in food products contributes to waste and resource depletion. Choose products with minimal packaging or those that use eco-friendly materials.
- **Join the clean plate club:** A significant portion of the food produced goes to waste, contributing to both environmental and economic losses. Plan meals, store food properly, and repurpose leftovers to minimize the amount of garbage you produce from cooking.
- **Buy local:** Do your part to reduce the carbon toll of food importation and trucking by eating things grown around you. Incorporate a variety of local, in-season fruits and vegetables into your diet.
- **Adopt plant-rich eating habits:** Plant-based diets have a lower environmental impact than diets centered

around animal products. Try to bring more plant-based meals into your routine.

- **Diversify your diet:** Change up your diet by enjoying a broader range of delicacies! The more you get used to eating local, seasonal vegetables in particular, the more you'll be doing to lessen your environmental impact. While it's tempting to buy packaged and prepared foods, once you shift your focus to eating more fresh produce, you'll be trying greens and root vegetables you haven't even heard of before.
- **Plant your own garden:** Growing your own produce, even on a small scale, reduces the distance your favorite herbs and vegetables travel to reach your plate. It can be a rewarding activity that can help you feel more connected to the environment.
- **Understand food as a process, not a product:** Consider the full journey your food takes from farm to table. Understanding the resources and efforts involved in agriculture and food production encourages more mindful consumption.
- **Make your voice heard:** Use your consumer power to drive demand for more sustainable and environmentally friendly food practices. Support businesses that prioritize ethical and eco-conscious choices.

ADVOCACY AND SPREADING AWARENESS

Along your path toward incorporating Eco Ripple principles of sustainable living, you're more influential than you might realize! Your role as a consumer extends beyond purchasing; it can also be a powerful driver of change. Your choices send a message that can really resonate with businesses, helping steer them toward more eco-conscious practices. By supporting environmentally responsible products and services, you're helping inform and reshape industries toward a greener future.

Becoming a More Climate-Conscious Consumer

Consumer influence doesn't stop at the cash register—it really has the potential to transform business strategies and objectives. The choices you make through exercising your buying power as a consumer can become a vote of confidence for companies that have demonstrated a commitment to sustainability. This, in turn, reinforces their efforts and helps encourage other businesses to follow suit. Your established preferences for eco-friendly options can help nurture a cycle of positive change.

Spreading Awareness Through Social Media and Personal Networks

Beyond the things you choose to buy, your voice holds immense sway. Social media and personal networks provide you with a platform to share insights, success stories, and practical tips for sustainable living. By utilizing these channels, you can magnify awareness, inspiring those around you to adopt eco-conscious habits and make informed choices themselves.

Conversations, whether with your neighbors or coworkers, can also be catalysts for change. Talking openly about climate change and sustainability amongst your peers can ignite both reflection and action. These types of dialogues help create a ripple effect of environmental consciousness in your local community, fostering a feedback loop of responsible living.

Your commitment to sustainability can serve as a way of bringing out positive behavior and habits in others. By advocating for eco-friendly practices and sharing your personal path toward eco-consciousness, you can encourage others to follow suit. When you take the time to cultivate a culture of awareness amongst those you interact with on a regular basis, you open up possibilities for a future where environmental stewardship becomes the norm.

As we wrap up this chapter, remember: Every individual choice that you make matters. Though it can lead to positive inner growth, embracing the principles of the Eco Ripple Effect isn't just about you; it's about being an agent of positive change for all of us. In the next chapter, we'll look at how your decisions hold the potential to help you define a sense of purpose in life and how you can create ripples and waves of positive impact that expand far beyond your immediate sphere of influence.

6

THE EARTH'S CUSTODIANS

I n a world brimming with challenges, our individual roles as custodians of the planet are far from trivial. The future of our planet rests not only on the actions of governments and big businesses but on the choices made by individuals like you and me. Behind our selfless actions lies the potential for empowerment, growth, and positive change. Sustainable living through embracing Eco Ripple principles holds an incredible potential to affect our own lives and the lives of those around us for the better.

A VISION OF EMPOWERED SUSTAINABILITY

We can envision a world of empowered sustainability, one where the potential for positive change lies within each of us. It's a positive vision that transcends mere existence, igniting a flame of hope and possibility.

What a Sustainable Future Might Look Like

Picture this: A world where lush rooftop gardens and vertical farms complement city skylines. Clean energy sources power bustling metropolises while wildlife thrives in carefully preserved pockets of nature. Streets are lined with cyclists, pedestrians, and electric vehicles. This is no distant dream; it's a tangible vision that can be realized through conscious choices and collective effort!

The Physical and Emotional Benefits

Beyond the transformation of landscapes, sustainable living offers tangible physical and emotional benefits. As you transition towards sustainable practices, you'll likely notice an increased appreciation for the interconnectedness of all living beings and the environment. This realization can lead to a greater sense of purpose, mindfulness, and empathy.

Aligning your lifestyle with sustainable principles fosters self-reflection and a deeper understanding of your values, ultimately contributing to personal growth and self-discovery. Through your investment in healing the wounds of the environment, you'll find that a profound self-healing potential simultaneously rises within you.

SUSTAINABILITY AS A JOURNEY OF PERSONAL GROWTH

Your eco-conscious choices have a far-reaching impact, not only on the environment but also on your own well-being. Embarking on the path of sustainability is not a mere adjustment of daily habits; it's a transformative journey that leads to self-discovery, deep insights, and a sense of purpose that resonates with the rhythm of the natural world.

With each conscious step you take to reduce your ecological footprint, you begin to perceive the world around you through a new lens—one that highlights the interconnectedness of all life forms. The sounds and sights of nature become poignant reminders of the delicate system of codependency and coexistence in which you play a vital part.

As you start to strip the excess consumption away from your life and daily habits, you may find a certain ease in the benefits that a more pared-down lifestyle can offer. Possessions that once held value to you seem less significant in the light of a life well-lived, unburdened by material excess. With each eco-friendly choice you make, you can further distill your own existence to its core, finding contentment in experiences and connections with others instead of possessions.

The Eco Ripple Effect can serve as a catalyst for personal growth, as it encourages you to align your actions with your value systems. The more you reduce waste, the more you can align yourself with a fundamental respect for nature's gifts. The act of mending, repurposing, or mindfully choosing items can help cultivate a deeper appreciation for craftsmanship, durability, and the stories that objects carry.

A profound sense of purpose can emerge when you consciously choose and try to walk this path. Understanding that your actions can contribute to the greater good can, in turn, fuel your determination to make a difference. Your once-individual ripples of change are then able to intermingle with a collective tide of positive impact, creating a wave that sweeps across communities and cultures. This sense of purpose can suddenly become the guiding force behind your choices, enriching your daily life with intention and meaning. By writing this very book, I myself have made my own little ripple; now you can make one, too!

You now have permission to shed the skin of indifference and adopt a renewed perspective on life—one that honors the symbioses of our fragile ecosystems, cherishes the bonds of close human connection, and values the authenticity found in purpose-driven living.

With each choice to reduce, reuse, and recycle, you send forth not only ripples of change but also echoes of inspiration that ripple out to others. Your family, friends, and community can all bear witness to your personal transformation, hopefully sparking curiosity and dialogue about the choices you've prioritized. As they witness your transformation into an eco-warrior, they too may even begin to perceive the beauty of your new, sustainable lifestyle.

In choosing the path of sustainable living that the Eco Ripple System offers, you can become a custodian of change, fostering a more harmonious coexistence between humanity and nature. The journey isn't just about reducing the amount of waste you produce; it's about expanding the horizons of possibility, deepening connections between yourself, others, and the earth, and, overall, contributing to the well-being of our planet and its inhabitants.

As you continue to tread lightly upon this Earth, remember that each step, each choice, creates a corresponding ripple that resonates far beyond your immediate actions. The journey of personal growth when you implement the Eco Ripple System can stand as a testament to the transformative power of conscious living—a journey that amplifies your sense of purpose, enriches your experiences, and creates a legacy of positive change for future generations.

A Pathway for Self-Discovery

Embarking on the path of sustainability isn't merely a change in daily habits; rather, it's a journey of personal growth and self-discovery. Each eco-conscious choice you make sends ripples of change through your life. As you consciously reduce your ecological footprint, you'll come to appreciate the richness of your surroundings, the value of simplicity, and the interconnectedness of all life forms. Through these experiences, you'll develop a deeper sense of purpose, aligning your actions with your core values. This alignment leads to a profound sense of fulfillment and a renewed perspective on life and what it means to be human.

THE RIPPLE EFFECT OF INDIVIDUAL ACTIONS

The more you reorient yourself toward Eco Ripple principles, the more you'll find that the actions of individuals have the astonishing power to spark significant change. In the following section, we'll take a look at the stories of those who dared to challenge the status quo with their disruptive climate action. Their tales resonate not just as accounts of personal achievement but as compelling examples of how the ripples of individual actions can amplify into waves of global transformation.

The common factor between these remarkable individuals is that each one of them harnessed their passions and convictions to foster change, not with grand gestures, but through the accumulation of conscientious choices. The ripple effect transcends borders, age, and backgrounds. It recognizes the universal truth that one person's determination can awaken the hearts and minds of many. When we look at stories of eco-warriors whose actions are reshaping the world, we can gain inspiration from the incredible journeys they undertook in their quest to transform their visions of climate activism into tangible realities.

The Power of The Ripple Effect

Your individual actions, no matter how seemingly small, can have profound and far-reaching effects. Remember: By making sustainable choices in your life, you can set a precedent that inspires others around you. Your actions create a ripple effect that echoes through your community and beyond, motivating others to join the movement for positive change.

Inspiring Stories of Eco-Awakening

Throughout history, some people have chosen to catalyze positive change through their climate actions. These climate champions demonstrate the power of an individual's commitment to environmental stewardship. Here are their stories:

Greta Thunberg

Greta Thunberg, a Swedish environmental activist, ignited a global youth movement for climate action. At the age of 15, Greta started playing hooky to protest outside of the parliament building in Stockholm, making an appeal for urgent climate action. Her solo strike gave birth to the Fridays for Future movement, inspiring millions of young people worldwide to rally for climate justice.

Her unwavering dedication has led to international fame and has even given her the opportunity to speak before world leaders, where she uses her platform to speak to the urgency of the climate crisis and the importance of holding global powers accountable for their actions.

Wangari Maathai

Kenyan environmentalist Wangari Maathai has left an indelible mark through her Green Belt Movement. Recognizing the interconnectedness of environmental degradation and poverty, she focused on tree planting, environmental conservation, and women's empowerment.

Her grassroots efforts led to the planting of over 30 million trees, contributing not only to reforestation but also to the empowerment of women and communities. Her legacy continues to inspire environmental movements across Africa and the world.

Julia Butterfly Hill

American environmental activist Julia Butterfly Hill is notable for having embarked on a two-year tree sit-in in a 1,000-year-old redwood tree named "Luna" to prevent its logging. Her act of courage brought attention to old-growth forest conservation and inspired a global movement to protect natural habitats.

Julia's steadfast commitment showcased the power of individual determination in safeguarding the environment, reminding us that our actions can indeed make a lasting impact.

Erin Brockovich

Erin Brockovich's relentless pursuit of justice revealed a corporate cover-up of groundwater contamination caused by toxic chemicals. Her efforts were instrumental in holding corporations accountable for their environmental impact, inspiring others to take action against environmental injustices.

Erin's story, which was even turned into a film where Julia Roberts starred as her, highlights the importance of uncovering and addressing hidden environmental threats while empowering communities to demand accountability.

Jadav Payeng

Jadav Payeng, known as the "Forest Man of India," transformed a barren sandbar into a lush forest ecosystem. Over several decades, he planted and nurtured countless trees, attracting diverse wildlife and rejuvenating the land.

His dedication to nature conservation highlights the profound impact of one individual's dedication and love for the environment. Jadav's story reminds us that even the most daunting environmental challenges can be tackled through persistence and care.

To Sum Things Up

These individuals, among others, have proven that change starts with a single individual. Their stories serve as a testament to the power of passion, determination, and unwavering commitment to environmental stewardship. Each of them embodies the spirit of being an Earth custodian, lighting the way for a more sustainable and harmonious world. As you read and think about their stories, remember that you, too, have the potential to create positive change, one action at a time.

EMBRACING HOPE AND TAKING ACTION!

The future of the Earth is shaped by your individual choices, both big and small. When you embrace the vision of empowered sustainability that the Eco Ripple System offers, where conscious living intertwines with personal growth and societal transformation, you're accepting the embrace of Mother Earth herself.

By understanding the ripple effect of your own actions, you can become an influential role model for others and be a catalyst for real, positive change. As this chapter concludes, know that the journey is never over. The path toward our choices matching our beliefs is a long one, but it's one worth taking. In the next chapter, we'll cover 101 steps that will help you embrace these principles with ease.

7

101 EASY STEPS

Welcome to 101 Easy Steps! You can use this chapter as a quick reference guide, utilizing its actionable and achievable steps to start making positive, eco-friendly changes in your life. From the small adjustments in your daily routine to the more conscious lifestyle decisions, these steps are designed to help you make more eco-conscious decisions while feeling like you're really making a difference. By incorporating these changes into your life, you'll not only reduce your own carbon footprint but will also be able to inspire others to join the movement for a greener future.

GROCERY SHOPPING

1. **Bring reusable bags:** When you head off to the supermarket, arm yourself with reusable bags. By

avoiding single-use plastics, you help reduce plastic pollution and conserve resources.

2. **Buy local produce:** Even when shopping at chain supermarkets, try to choose locally-grown fruits and vegetables over imported ones.

3. **Support farmers' markets and community-supported agriculture (CSA) programs:** These programs help distribute fresher, locally-grown produce in your area! Supporting nearby farmers not only boosts your local economy but also helps cut down on long-distance trucking.

4. **Choose minimal packaging:** Buy food products with minimal or recyclable packaging. Or better yet:

5. **Buy things with no packaging at all:** the skin of fruits and vegetables is nature's own packaging—we don't need them to be wrapped in plastic! This simple choice contributes to reducing waste and curbing the excessive use of resources.

6. **Shop in bulk:** When possible, buy food items in bulk. Make sure that the bulk items you're buying aren't individually wrapped, though. This practice not only decreases the amount of packaging but also saves you money in the long run.

7. **Opt for reusable containers for liquid soaps, detergents, and cleaning products:** Swap out disposable containers for reusable ones and buy from natural bulk stores. This easy change diminishes the demand for single-use plastics and fosters a more sustainable kitchen.

8. **Purchase locally sourced honey:** Buying your honey locally supports local beekeepers, keeps local pollinating bee colonies in your neighboring ecosystems, and reduces your carbon footprint as opposed to buying commercial honey that's trucked across the country.

9. **Avoid buying bottled and canned beverages:** Bottled water, soda, and juices mean more energy expenditure in processing your recycling.

10. **Bring your own containers:** If you're the type that can't avoid the grocery store steam table or prepared foods section, this helps you reduce single-use plastic waste. Just make sure they've been cleaned thoroughly.

AROUND THE HOUSE

1. **Switch to LED light bulbs:** Replace traditional bulbs with energy-efficient LED ones. These bulbs consume less energy and last longer, ultimately reducing your carbon footprint. Incandescent and halogen bulls have been largely phased out throughout large swaths of the world. If you're still using old light bulbs—get a clue!

2. **Unplug electronic devices:** When devices aren't in use, unplug them to prevent "phantom" energy consumption. While difficult in practice, this small action adds to significant energy savings. Let's be honest: Do I really unplug my TV every time I leave the house? Absolutely not! But the more we're aware of phantom power drain, the less energy we consume.

3. **Use natural cleaning products:** Switch to natural cleaning products like white vinegar. This change minimizes chemical exposure, protecting both your health and the environment. The thing to remember about any cleaning product is that it eventually ends up down the drain and may harm local ecosystems.

4. **Install a programmable smart thermostat:** Control your home's temperature efficiently by installing a programmable thermostat. This step helps you save energy and reduce your utility bills. Make sure to implement a smart energy program for times that you're at work, and when you're on vacation, make sure to program a conservative energy mode.

5. **Set up a composting system:** Create a composting system for food scraps and garden waste. Composting helps cut down on landfill waste while also producing nutrient-rich soil. You can either arrange curbside pickup or use the compost in your own garden.

6. **Insulate your home:** Proper insulation keeps your home comfortable while conserving energy. This contributes to lower utility bills and a reduced carbon footprint. Especially if you live in an older home, it's worth taking a look at sealing up those leaks, cracks, and drafty walls and windows.

7. **Plant native trees and plants:** Enhance your surroundings with native trees and plants. They support local ecosystems and provide habitat for wildlife. You're not only helping produce more oxygen this way but will also have the satisfaction of connecting yourself with the natural world.

8. **Purchase energy-efficient appliances:** When buying new appliances, choose energy-efficient models. These appliances use less energy, benefiting both the environment and your wallet. Check online reviews and read up on your Energy Star certifications.

9. **Plant a garden to filter and manage rainwater runoff naturally:** This can be especially useful if you live in a low-lying area already prone to flooding.

10. **Install solar panels on your roof:** This can help your household harness renewable energy while also lowering your electricity bills. I did it at my house, and I can't even imagine what life was like before it!

11. **Repair and repurpose:** Instead of buying new ones, repair and repurpose old furniture and home decor items. Even if you're not handy, with nontoxic wood glue, clamps, and some patience, you can do a lot!

12. **Create a dedicated space for recycling in each room:** This will help encourage the members of your household to use proper recycling practices, even if they're the type to carry snacks and beverages around the house.

KITCHEN

1. **Reduce food waste:** Plan your meals and use leftovers wisely to minimize food waste. This practice saves money and reduces the environmental impact of food production. Don't cook more than you need to feed your family. While it's great to have a massive feast once

in a while, cooking smaller quantities will mean less overall waste.

2. **Reduce food waste further by composting kitchen scraps:** From coffee grounds to eggshells, the Earth will thank you for composting your scraps. You can use it in your own garden, and in some areas, you may even be able to put it out for curbside pickup along with your household recycling.

3. **Use cloth napkins and only reusable dishware:** Make the switch to cloth napkins and reusable dishware to avoid disposable alternatives. This small change can make a huge difference. While paper towels can be convenient, using cloth rags is a much more eco-friendly way to clean up spills.

4. **Ditch single-use plastics:** Say goodbye to single-use plastic utensils and paper plates. If you go to any dollar stores these days, you can find reusable bamboo plates and utensils that are great for picnics, barbecues, and other outdoor events. While they won't last as long as your at-home dishware, you won't have to throw them out after just one use.

5. **Grow your own herbs and vegetables**: Cultivate your own herbs and vegetables at home. This step promotes sustainability, self-sufficiency, and a deeper connection with your food. You'll also find that it can cut down on grocery bills.

6. **Purchase locally produced and organic food:** Support local farmers and choose organic products. This choice reduces the use of harmful chemicals, maintains vital ecosystems, and supports sustainable agricultural

practices. You'll also be cutting down on the carbon imprint of food transport, as the produce will be going through fewer hands before landing in your kitchen.

7. **Replace disposable coffee filters with reusable ones made from cloth or metal:** This small change can save a significant amount of paper waste if you're a regular coffee drinker. You can also use reusable tea strainers versus paper tea bags.

8. **Use a water-saving dishwasher to reduce water consumption during dishwashing:** Hand-washing wastes more water, so make sure to use the eco setting on your dishwasher if this option is available.

9. **Say goodbye to crystal glassware and fine china:** While it's nice to have family heirlooms around, delicate glassware and dishware that needs to be hand-washed wastes water. Stick with your regular plates and flatware that are dishwasher-safe.

10. **Use dishwasher-safe pots and pans:** Reduce water use by machine-washing pots and pans you'd normally hand-wash. Stick to nontoxic, nonstick surfaces for faster, easier cleanup.

11. **Purchase fair trade and ethically sourced coffee and tea products:** Look for growers that utilize less water resources and use sustainable land management and agricultural practices.

12. **Avoid flimsy plastic food storage containers that can crack or chip:** While glass uses more energy to produce, glass storage containers with plastic lids can last a lifetime if treated with care.

BATHROOM

1. **Choose cruelty-free and environmentally friendly beauty products:** When selecting beauty products, check labeling for cruelty-free and eco-friendly certifications. Using products like these helps support ethical practices, reduces overall environmental harm, and doesn't rely on animal testing.

2. **Make your own skincare products:** Experiment with creating your own skincare products using natural ingredients. Ingredients like coconut oil and fresh herbs from your garden can be used to create natural cosmetics. This practice can help minimize your exposure to synthetic chemicals and reduce packaging waste.

3. **Use refillable containers:** Replace single-use product containers with refillable ones. This practice cuts down on plastic waste and encourages a more sustainable bathroom routine. While it's easy to throw away items that have been sitting around in the bathroom for sanitary reasons, with proper cleaning practices, you can easily keep disposable items like cotton swabs in glazed ceramic jars instead of disposable plastic containers without getting grossed out. When I travel, I reuse face lotion jars that fit TSA carry-on guidelines (under 4 oz) as shampoo containers. This keeps me from buying sample-size travel toiletry items.

4. **Avoid products with microbeads:** Microbeads, which are used in many body washes, soaps, and scrubs, can

harm aquatic life. This simple choice to avoid using them can help protect marine ecosystems. There are plenty of facial scrubs and other products that use natural abrasives instead of microbeads, so consider using one of those products instead.

5. **Reduce water consumption:** Take shorter showers and turn off the tap while washing your face or brushing your teeth. This reduces water waste and conserves this precious resource.

6. **Install low-flow faucets and showerheads:** Installing these can help your household conserve water without sacrificing pressure.

7. **Replace conventional toilet paper with eco-friendly, recycled options:** Lightweight single-ply recycled toilet paper may be off-putting to some, but it can help you use less versus bulky double-ply rolls.

8. **Use a bidet attachment or bidet toilet seat to reduce toilet paper usage:** These handy devices, many of which can be added to your existing toilet without any additional plumbing, can reduce your toilet paper usage substantially.

9. **Purchase bamboo toothbrushes:** Bamboo is a sustainable alternative to plastic toothbrushes. Considered a renewable resource, Bamboo spreads rapidly and is a lightweight material that requires minimal processing.

10. **Choose natural and biodegradable personal hygiene products:** Help out the environment by minimizing emissions and pollution by avoiding petroleum-based plastic products.

WARDROBE

1. **Avoid fast fashion and opt for high-quality, timeless clothing pieces:** Avoid trends and opt for quality and durability. It'll make a huge difference over time! While buying these items may be more expensive, it'll help minimize your need to shop and dispose of unused items.

2. **Organize clothing swap events with friends or in your community:** Small, local events like these can have a lesser carbon impact than chain thrift stores, which generally have larger energy expenditure due to larger quantities of clothes and home items donated.

3. **Purchase second-hand clothing and accessories:** Smaller, independent thrift stores and secondhand shops can be great choices for scoring clothing and accessories at lower prices. This reduces textile waste and helps support a circular economy.

4. **Choose sustainable and ethically produced clothing:** Prioritize sustainable and ethically produced clothing brands and materials. Learning about where clothes are produced can help you make smarter decisions about which brands to buy from. Buying from brands that use responsible manufacturing practices and source material and labor domestically can help reduce the negative impact of fast fashion.

5. **Donate or recycle old clothing:** Instead of throwing away old articles of clothing, donate or recycle them.

This keeps textiles out of landfills and provides affordable fashion for others.

6. **Upcycle old clothing:** Use your old clothes to make new things. Old shirts can be woven into rugs, towels, and pot grips, while old ripped jeans and other heavy natural-fiber wovens can be used to make durable, stuffed dog toys for your pooch.

7. **Mend and repair clothes:** When clothes tear or wear out, repair them instead of replacing them. This extends the lifespan of your garments and reduces consumption. An experienced tailor or cobbler can do wonders for your worn-out clothing items and accessories.

8. **Learn how to fix and make your own clothes:** Start by learning basic clothing repair skills like hand-sewing, sewing buttons, and patching small holes. Then, you'll be ready to move on to basic garment construction techniques like following patterns and machine sewing. Soon enough, you'll be doing your own tailoring and repairs at home, and once you get good at it—you can even start making your own unique fashion statements with custom-created garments.

LIFESTYLE

1. **Reduce household clutter:** Prioritize quality over quantity in your belongings. Reducing excess clutter around the house helps reduce consumption and encourages more mindful, meaningful living.

2. **Choose experiences over possessions:** Shift your focus towards experiences and memories rather than material possessions. This practice can enrich your life while also decreasing your ecological footprint.

3. **Reduce meat consumption:** Incorporate more plant-based meals into your diet to reduce meat consumption. Plant-based diets have a lower environmental impact, and they're healthier. You don't have to forgo meat completely! But take a cue from diets like the Mediterranean diet, and you'll find that you can eat healthier and more carbon-neutral without sacrificing flavor, fats, and all the things you enjoy.

4. **Support sustainable and ethical brands:** When making larger purchases, like cars, opt for brands that prioritize sustainability and ethical practices. This helps cast your vote for responsible business practices and the eco-revolution.

5. **Engage in community activities:** Participate in community activities that promote environmental awareness. Engaging with like-minded individuals helps foster a mutual sense of belonging that can drive positive change.

6. **Practice mindfulness and meditation:** This can help you cultivate a deeper connection with your surroundings and make you feel more connected to the environment.

7. **Purchase carbon offsets and make green investment decisions:** Carbon offsets may help address remaining emissions from your daily activities, while a portfolio of

soundly invested equity in companies with good ESG figures can go a long way.

8. **Engage in more outdoor activities on your vacations:** Things like hiking and camping can help foster a greater appreciation for nature. It is also less carbon-intensive than going to theme parks and other big attractions, staying in hotels, or going on cruises.

9. **Volunteer for local school boards and business organizations:** Encourage local schools and businesses to adopt eco-friendly practices and sustainability initiatives. Your voice can make a difference.

10. **Share your knowledge and passion for sustainability with others:** This can help to inspire positive change in your community. You can consider holding workshops, seminars, nature walks, or trash pick-up events.

11. **Be an eco-friendly influencer:** Encourage friends and family members who aren't so green to adopt eco-friendly practices in their own lives. Your enthusiasm around these issues can even inspire positive change within your social circles. That's what the Eco Ripple Effect is all about!

TRANSPORTATION

1. **Walk, bike, or carpool:** Opt for walking, biking, or carpooling for short trips. These alternatives reduce emissions and promote healthier, more carbon-neutral modes of transportation.

2. **Use public transportation:** Choose public transportation whenever possible to reduce your carbon footprint. This also helps by reducing traffic congestion and air pollution.

3. **Choose electric or hybrid vehicles:** If you're in the market for a new vehicle, consider electric or hybrid options. These vehicles produce fewer emissions and can help contribute to cleaner air.

4. **Maintain your vehicle:** Regular vehicle maintenance, such as changing your oil and checking your tire pressure, ensures optimal fuel efficiency. These things can make a huge difference if you're still driving an older conventional gas or diesel car, giving you the power to make meaningful change, even if you can't afford a newer car.

5. **Combine errands:** Plan your trips strategically to combine errands. This reduces unnecessary travel and saves time and fuel. Using GPS navigation apps like Google Maps and Waze are great ways to avoid traffic and find the lowest-mileage routes, helping you cut down on emissions while saving at the gas pump.

6. **Don't fly:** Reduce the amount of flights you take per year. Consider other options, such as traveling by train whenever possible. Reduce the frequency of business and international travel whenever possible.

OFFICE/BUSINESS

1. **Print double-sided on low-weight recycled paper:**
   With double-sided printing, you'll use half the amount
   of paper! Lighter-weight, recycled paper carries a much
   lower carbon footprint. Printing in black and white vs.
   color uses less ink per page. Make sure the settings in
   your print dialog are set to "black and white" and not
   "grayscale," as this setting can use extra ink. Simple
   choices like these can help conserve resources if you
   still need to print out documents at your place of work.

2. **Use digital documents and emails:** Minimize paper
   usage by utilizing digital documents and emails. This
   reduces the need for printed materials. Today, it's even
   possible to sign contracts online. If your job requires
   you to review lengthy contracts or thumb through
   bulky packets or booklets, ask if there's a digital version
   available.

3. **Provide recycling bins:** Make recycling easy by
   utilizing recycling bins in the office. Encourage proper
   waste disposal practices and less usage of printed
   documents among colleagues.

4. **Implement energy-saving practices:** Turn off lights
   and computers when not in use. Implementing energy-
   saving practices in the workplace conserves energy and
   reduces business overhead.

5. **Work from home:** Remote work options can be a great
   way to reduce your carbon footprint if your employer
   allows it. Working from home cuts down on emissions

from long commutes while also helping support work-life balance.

6. **Limit travel by car:** Whether it's for sales meetings or corporate summits, many of the discussions that happen in key meetings can be done over Zoom rather than in person.

7. **Optimize service and delivery fleet routes:** If your business requires the use of one or multiple vehicles, make sure to implement smart route-planning practices that minimize fuel waste. This will save your business money while helping out the Earth.

8. **Reevaluate power generation HVAC and utility systems:** As some companies rely on their own power plants and generators, and almost all rely on HVAC systems, it's wise to review these systems periodically to see if they need updating for better efficiency and less carbon output.

9. **Move to a better, more modern office building:** If your business rents office space in an old building filled with cracks and leaks, in effect, you're contributing to the release of carbon gasses. Consider relocating your business to a LEEDS-certified building.

10. **Reevaluate your IT infrastructure:** Servers and computer towers that remain powered even outside of business hours can be a big drain on electricity. If something's not necessary to leave powered on, shut it down before you head out for the day.

TECHNOLOGY

1. **Don't get a new phone every year:** While cellular companies incentivize swapping out and constantly upgrading mobile devices, try to stick with the one you have for a number of years before trading it in to be refurbished and resold or safely recycled.

2. **Reduce screen time:** Decrease your screen time to save energy and promote well-being. This practice conserves device battery life and fosters a healthier lifestyle. The more hours we spend online, the more our carbon footprint widens.

3. **Reduce streaming time:** Did you know that all that binge-watching exerts a carbon toll? While vegging out for hours with your favorite TV shows, movies, and YouTube videos can be fun, it's also highly energy-intensive.

4. **Opt for digital media and e-books:** Choose digital media and e-books over printed materials. This reduces paper usage and supports a more sustainable media consumption habit. That's, after all, why you chose to buy this book, right?

5. **Use power-saving settings:** Enable power-saving and standby settings on your electronic devices. This conserves energy and also can help extend the battery life of your devices.

6. **Turn off and unplug desktop computers when not in use:** Though it may seem obvious, many forget to do

this and leave their computer on sleep for weeks at a time, even if the computer isn't in use.

7. **Choose energy-efficient desktop computers that don't exceed your needs:** When upgrading or purchasing new computers or peripherals, opt for energy-efficient models appropriate for the tasks you need them for. Unless you're using your computer for video editing, advanced graphics, or gaming, you really don't need a top-of-the-line, resource-intensive power-leeching model.

8. **Buy an energy-efficient laptop, Chromebook, or tablet for on-the-go computing:** Smaller, less resource-intensive devices should be used for daily computing tasks like emails, internet browsing, and word processing.

9. **Check the energy specs on TVs and home theater equipment:** Newer models are often more energy-efficient, saving you money and helping reduce your household's electricity consumption.

10. **Recycle old electronics (when possible):** Dispose of old electronics responsibly to prevent e-waste. Many electronics contain hazardous materials that can harm the environment if not recycled correctly. Always check the labeling to make sure you're in compliance, and ask your local municipality about technology drop-off days where you can safely recycle items like old computers and other devices.

GIFTING

1. **Give experiences:** Instead of physical items, consider giving experiences, like a gift certificate to go to a spa or a reservation for a glamping trip. This type of giving can be more thoughtful than just buying something from a store and can also be a more carbon-neutral way to express your gratitude as opposed to buying a manufactured product.

2. **Give homemade gifts:** Whether it's a tin of warm cookies or a hand-knit sweater, homemade gifts can be even better than cheap merchandise. Focus on the things you know they'll like, and you'll find that this can be a great way to save money while giving something unique and special.

3. **Wrap gifts with eco-friendly materials:** When wrapping gifts, use reusable fabric or newspaper instead of disposable wrapping paper. This reduces waste and adds a fun, personalized touch.

4. **Choose sustainable and ethical brands:** Prioritize sustainable and ethical brands when purchasing gift items. This ensures that the presents you give to others align with your own values.

5. **Support charities and nonprofits:** Consider donating to charities and nonprofits as gifts in someone's name. This supports meaningful causes and spreads positivity without relying on the resource-heavy manufacturing and shipping processes of physical items.

6. **Give quality items:** Instead of buying cheaply manufactured products that can easily break, consider gifting higher-quality items that can last a lifetime.

BABY ITEMS

1. **Rely on natural, nonpetroleum-based oils and ointments:** While diaper rash can be a real pain for your young one, make sure to buy products that don't rely on petroleum extraction and the associated chemical processing of crude oil.

2. **Buy cloth diapers:** Opt for cloth diapers instead of disposable ones. While it may seem disgusting to you at first, cloth diapers are reusable and reduce the amount of waste generated.

3. **Opt for eco-friendly baby toys:** Choose eco-friendly and organic baby products. These products are better for your baby's health and the environment. While all baby products are free of harsh chemicals, big manufacturers still load many of their baby products with nonrecyclable plastics that are produced with carbon-intensive processes and cause damage to the environment.

4. **Use hand-me-downs or borrow baby items:** Reduce consumption by using hand-me-downs or borrowing baby items from friends or relatives. This can help reduce the demand for new products while keeping landfills from overflowing with baby clothes that don't fit anymore.

5. **Create a sustainable baby registry:** When creating a baby registry, prioritize sustainable and nontoxic items. This ensures a greener start for your little one, even if some of your friends and family aren't yet clued in.

6. **Repurpose baby gear:** Instead of buying new items, repurpose baby gear for different stages. This reduces waste and saves money in the long run. You can alter baby clothes, and there are even baby clothing lines that produce clothes designed to expand as your child gets bigger.

These are just some of the easy steps you can take to lead a more sustainable lifestyle. By incorporating these changes into your daily routine, you'll contribute to a greener future and inspire others to join you on this journey. Remember, every small effort adds up to make a significant impact on the planet!

I would love to hear from you!

If you found value in this book, I kindly ask for your review and for you to share your experience with others. Your feedback helps us spread the message of sustainability to even more individuals.
It's through your support and reviews that my book is able to reach the hands of other readers.

Please take 60 seconds to kindly leave a review on Amazon. Please scan the QR code below. If you reside in a country that is not listed, please use the link provided in your order.

All it takes is 60 seconds to make a difference!

CONCLUSION

As you've flipped through the pages of this book, you've made the conscious choice to set out on a path of discovery and empowerment, uncovering the incredible potential within each of us to become stewards of our planet. The essence of sustainable living lies not in grand gestures but in the everyday choices we make. These choices, when compounded, can lead to real, positive change and a brighter, more resilient future.

At its core, this book is a call to action, an invitation to embrace your role as an Earth custodian through the Eco Ripple Effect principles. It's a reminder that the seemingly small decisions you make each day—what you consume, how you travel, and the products you choose—have profound implications that extend far beyond your immediate surroundings. It's my attempt to send out a small ripple to you.

Though I still haven't been able to find a conclusive figure on the climate impact of my own daily commuting choices, I became so involved in researching it that I ended up writing this book.

The key takeaway is crystal clear: Your actions matter! You possess the power to initiate a ripple effect of positive change that transcends borders, generations, and cultures. By adopting sustainable practices, you're not only reducing your carbon footprint but also potentially igniting a spark that can inspire others to do the same.

I want to invite you to join me in the near future in a world where sustainability is more than a buzzword but a way of life —a world where communities thrive, where harmony between humanity and nature is the norm, and where the legacy we leave for future generations is one of hope and promise. This vision is within reach, and you are a vital part of its realization.

As you close this book, let the message resonate: You're never just an observer of the world but a creator of its future. Seize this moment to reflect on the steps you've learned, the principles you've embraced, and the inspiration you've gained. Let that reflection guide you in the days, weeks, and years to come.

Now, as you stand at the threshold of this new awareness, consider your circle of influence. Share your journey with others, for within your stories and actions lies the ability to inspire. Engage in conversations that ignite curiosity and passion for sustainable living. Celebrate your successes and challenges, for they're a testament to your commitment.

The path to a greener world is not always linear, and change may seem incremental at times. But remember that each choice you make sends ripples far and wide. So whether it's reducing your carbon footprint, advocating for policy changes, or simply living mindfully, know that you are part of a collective movement for change.

Now, dear reader, I encourage you to put the knowledge you've gained into practice. Be a catalyst for the shift we so urgently need. Embrace the journey of sustainable living not as a burden but as a privilege. Let your actions echo through time, inspiring generations to come. Your choices have the potential to shape the course of history and create a legacy that future Earth custodians will be proud to inherit.

Thank you for choosing to heighten your eco-awareness with me. If you found value in this book, we kindly ask for your review and for you to share your experience with others. Your feedback helps us spread the message of sustainability to even more individuals. If you invested in a paper copy, please give it to a friend or coworker or leave it in the office pantry or a local book exchange. You have an opportunity to share the message without any incremental CO2e emission.

Remember: The world is changed by those who believe they can change it. Your actions matter, and they matter now. Let's continue to make the world a better place, one light-on-the-earth treading, sustainable step at a time.

REFERENCES

Acteevism. (2023). *100 ways to live more sustainably – simple sustainable living actions*. Acteevism.com. https://www.acteevism.com/2019/11/11/100-ways-to-live-more-sustainably

AGL. (2023). *Benefits of reducing your carbon footprint*. Agl.com.au. https://www.agl.com.au/discover/sustainability/benefits-reducing-carbon-footprint

Anderson, K. (2018, December 11). *What is background extinction rate and how is it calculated?* Population Education. https://populationeducation.org/what-is-background-extinction-rate-how-is-it-calculated

Antos, D. (2020, November 9). *Choose plastic over glass for better sustainability*. Drug Plastics & Glass Co., Inc. https://www.drugplastics.com/choose-plastic-over-glass-for-better-sustainability/

Arango, T., & Healy, J. (2023, August 23). *With reported missing in Maui topping 1,000, officials will release names*. The New York Times. https://www.nytimes.com/2023/08/23/us/maui-missing.html

ASM Metal Recycling. (2022, January 27). *The effects of reduced aluminium can use*. ASM Metal Recycling. https://www.asm-recycling.co.uk/blog/the-economic-effects-of-reduced-aluminium-can-use/

Atalla, G., Mills, M., & McQueen, J. (2021, December 6). *Six ways that governments can drive the green transition*. Ey.com; EY. https://www.ey.com/en_gl/government-public-sector/six-ways-that-governments-can-drive-the-green-transition

BCG. (n.d.). *How companies can decarbonise their supply chains*. Bcg.ft.com. https://bcg.ft.com/article/companies-decarbonise-their-supply-chains

Bee Informed. (2023, June 22). *United states honey bee colony losses 2022–23: Preliminary results from the bee informed partnership*. Beeinformed.org. https://beeinformed.org/2023/06/22/united-states-honey-bee-colony-losses-2022-23-preliminary-results-from-the-bee-informed-partnership/

Begum, T. (2021, May 19). *What is mass extinction and are we facing a sixth one?* Natural History Museum, London. https://www.nhm.ac.uk/discover/what-is-mass-extinction-and-are-we-facing-a-sixth-one.html

Benefits of nature reserves. (2021, February 11). Https://Norseywood.org.uk/. https://norseywood.org.uk/benefits-of-nature-reserves

Biodiversity Conservation Trust. (2023). *What is biodiversity and why is it important?* Www.bct.nsw.gov.au. https://www.bct.nsw.gov.au/what-biodiversity-and-why-it-important

Biology Online. (2019, November 30). *Ecological niche definition and examples.* Biology Articles, Tutorials & Dictionary Online. https://www.biologyonline.com/dictionary/ecological-niche

Bisset, V. (2023, July 29). *The U.N. warns "an era of global boiling" has started. What does that mean?* Washington Post. https://www.washingtonpost.com/climate-environment/2023/07/29/un-what-is-global-boiling/

Boren, Z., Jordan, L., Ross, A., Wasley, A., & Mendonça, E. (2021, October 13). *British supermarket cheese linked to catastrophic deforestation in brazil.* Unearthed. https://unearthed.greenpeace.org/2021/10/13/cargill-deforestation-brazil-uk-dairy-cadbury-cheese/

Can consumers drive corporate sustainability? (n.d.). Www.greenbiz.com. https://www.greenbiz.com/article/can-consumers-drive-corporate-sustainability

Carbon capture and utilization. (2023, August 30). In Wikipedia. https://en.wikipedia.org/w/index.php?title=Carbon_capture_and_utilization&oldid=1173006652

Chen, J. (2022, July 20). *Impact investing definition.* Investopedia. https://www.investopedia.com/terms/i/impact-investing.asp

Cho, R. (2019, June 20). *How climate change impacts the economy.* State of the Planet. https://news.climate.columbia.edu/2019/06/20/climate-change-economy-impacts

Cho, R. (2020, September 22). *Why climate change is an environmental justice issue.* Columbia Climate School: State of the Planet. https://news.climate.columbia.edu/2020/09/22/climate-change-environmental-justice

Chung, R. (2023, March 2). *Why carbon accounting is so important in this day and age.* Cundall. https://www.cundall.com/ideas/blog/why-carbon-accounting-is-so-important-in-this-day-and-age

Clarke, R. (2021, August 17). *Fast fashion's carbon footprint.* The Carbon Literacy Project; The Carbon Literacy Trust. https://carbonliteracy.com/fast-fashions-carbon-footprint/

ClientEarth Communications. (2022, February 18). *Fossil fuels and climate change: The facts.* Www.clientearth.org. https://www.clientearth.org/latest/latest-updates/stories/fossil-fuels-and-climate-change-the-facts

ClimateScience. (2022, September 12). *Does personal action matter?* Climatescience.org. https://climatescience.org/advanced-personal-action

CNN, B. A. F. (2020, December 9). *10 surprising sources of greenhouse gasses.* CNN. https://edition.cnn.com/2019/06/03/world/gallery/surprising-sources-greenhouse-gas-emissions-intl/index.html

Concern Worldwide US. (2022, May 3). *The 7 human activities that cause climate change.* Concernusa.org. https://concernusa.org/news/human-activities-that-cause-climate-change

Consumers can positively influence business behaviour. (n.d.). ReGenerate. https://www.re-generate.org/blog-content/consumers-can-positively-influence-business-behaviour

Corporate social responsibility: environmental impact. (2018, July 3). Nibusinessinfo.co.uk. https://www.nibusinessinfo.co.uk/content/corporate-social-responsibility-environmental-impact

Council on Foreign Relations. (2023). *How do governments combat climate change?* World101 from the Council on Foreign Relations. https://world101.cfr.org/global-era-issues/climate-change/how-do-governments-combat-climate-change

Creel, T. (2020). *How corporate social responsibility influences brand equity.* Imanet.org. https://www.imanet.org/-/media/5ab6966eedf6461ab37f f8b8f2fb6eba.ash

Deloitte. (2014). *The-growing-power-of-consumers: A Deloitte insight report.* https://www2.deloitte.com/content/dam/Deloitte/uk/Documents/consumer-business/consumer-review-8-the-growing-power-of-consumers.pdf

Deloitte. (2022, May 23). *Deloitte research reveals inaction on climate change could cost the world's economy US$178 trillion by 2070.* Www.deloitte.com. https://www.deloitte.com/global/en/about/press-room/deloitte-research-reveals-inaction-on-climate-change-could-cost-the-world-economy-us-dollar-178-trillion-by-2070.html

Delubac, A. (2023). *5 best carbon footprint calculators.* Greenly.earth. https://greenly.earth/en-us/blog/company-guide/5-best-carbon-footprint-calculators

Denchak, M. (2019, July 16). *Greenhouse effect 101.* NRDC; Natural Resources Defense Council. https://www.nrdc.org/stories/greenhouse-effect-101

Denchak, M. (2022, June 1). *Fossil fuels: The dirty facts.* NRDC. https://www.nrdc.org/stories/fossil-fuels-dirty-facts

Denchak, M., & Turrentine, J. (2021, September 1). *What is climate change?* NRDC. https://www.nrdc.org/stories/what-climate-change

Denton, M., & Perrella, J. (2021). *How US Banks Are Addressing Climate Risk and*

Sustainability. Www.moodysanalytics.com. https://www.moodysanalytics. com/articles/2021/how-us-banks-are-addressing-climate-risk

Direct Energy. (n.d.). *25 energy efficiency tips for your home.* Www.directenergy.- com. https://www.directenergy.com/learning-center/25-energy-efficiency- tips

Dorian, M., Gorin, T., Yamada, H., & Yang, A. (2021, June 10). *Erin Brockovich: The real story of the town three decades later.* ABC News. https://abcnews.go. com/US/erin-brockovich-real-story-town-decades/story?id=78180219

Earth Eclipse. (n.d.). *What are national parks? Why are national parks important and how do they help the environment?* Eartheclipse.com. https://eartheclipse.com/ environment/national-parks.html

Eartheasy. (n.d.). *45+ ways to conserve water in the home and yard.* Eartheasy Guides & Articles. Retrieved August 9, 2023, from https://learn.eartheasy. com/guides/45-ways-to-conserve-water-in-the-home-and-yard

Editors of Encyclopaedia Britannica. (2020). *Extinction rate.* In Encyclopædia Britannica. https://www.britannica.com/science/extinction-rate

Enel Green Power. (2023, February 28). *OK, so it's bad, but not that bad: In spite of the climate crisis, there's still plenty of good news.* Www.enelgreenpower.com. https://www.enelgreenpower.com/learning-hub/gigawhat/search-articles/ articles/2023/02/climate-crisis-opportunities

Energy Education Canada. (n.d.). *Species migration.* Energyeducation.ca. https:// energyeducation.ca/encyclopedia/Species_migration

Environmental and Energy Study Institute. (2021, July 22). *Fossil fuels.* Www.eesi.org. https://www.eesi.org/topics/fossil-fuels/description/

EPA. (2017, September 7). *Frequent questions on recycling.* Www.epa.gov. https:// www.epa.gov/recycle/frequent-questions-recycling#recycling101

EPA. (2018, November 7). *How do I recycle?: Common recyclables.* Www.epa.gov. https://www.epa.gov/recycle/how-do-i-recycle-common-recyclables

EPA. (2022, October 19). *Climate change impacts on agriculture and food supply.* Www.epa.gov. https://www.epa.gov/climateimpacts/climate-change- impacts-agriculture-and-food-supply

Euractiv. (2005, February 16). *The great divide: EU-US approaches to climate change.* Www.euractiv.com. https://www.euractiv.com/section/sustainable- dev/news/the-great-divide-eu-us-approaches-to-climate-change

European Parliament. (2018, August 3). *Reducing carbon emissions: EU targets and policies.* Www.europarl.europa.eu. https://www.europarl.europa.eu/news/

en/headlines/society/20180305STO99003/reducing-carbon-emissions-eu-targets-and-policies

European Parliament Think Tank. (2016, November 19). *Climate policies in the EU and USA: Different approaches, convergent outcomes?* | Think Tank | European Parliament. Www.europarl.europa.eu. https://www.europarl.europa.eu/thinktank/en/document/EPRS_BRI(2015)571347

Ferguson, L. (2019, May 21). The extinction crisis. Now.tufts.edu. https://now.tufts.edu/2019/05/21/extinction-crisis

Gavin, M. (2019, June 6). *5 corporate social responsibility examples that were successful.* Harvard Business School. https://online.hbs.edu/blog/post/corporate-social-responsibility-examples

Goldman Prize. (2021, September 1). *How grassroots environmental activism has changed the course of history.* Goldman Environmental Prize. https://www.goldmanprize.org/blog/grassroots-environmental-activism

Gray, E. (n.d.). *Global climate change impact on crops expected within 10 years, NASA study finds.* Climate.NASA.gov. https://climate.nasa.gov/news/3124/global-climate-change-impact-on-crops-expected-within-10-years-nasa-study-finds

Gray, R. (2019, March 4). *Sixth mass extinction could destroy life as we know it–biodiversity expert.* Ec.europa.eu. https://ec.europa.eu/research-and-innovation/en/horizon-magazine/sixth-mass-extinction-could-destroy-life-we-know-it-biodiversity-expert

Greeka. (n.d.). *Environmental policies of Greek ferry companies.* Greeka.com. https://www.greeka.com/travel-services/greek-ferries/environmental-policies/

Greenwashing: What is it, why is it a problem, and how to avoid it. (2021, July 23). Earth.org - Past | Present | Future. https://earth.org/what-is-greenwashing

Greenwood, S. (2021, September 14). *In response to climate change, citizens in advanced economies are willing to alter how they live and work.* Pew Research Center's Global Attitudes Project. https://www.pewresearch.org/global/2021/09/14/in-response-to-climate-change-citizens-in-advanced-economies-are-willing-to-alter-how-they-live-and-work

HomeWater. (2022, July). *Water conservation: Why it's important and what you can do.* HomeWater. https://www.homewater.com/blog/water-conservation-why-its-important

Hosokawa, R., & KIdera, M. (2022, May 2). *Climate refugees: The world's silent crisis.* Nikkei Asia. https://asia.nikkei.com/Spotlight/Datawatch/Climate-refugees-The-world-s-silent-crisis

How does energy efficiency help the environment? (n.d.). Www.energysage.com. https://www.energysage.com/energy-efficiency/why-conserve-energy/environmental-impact-of-ee

How to reduce waste: 21 ideas for zero waste living. (2022, November 10). One Tree Planted. https://onetreeplanted.org/blogs/stories/how-to-reduce-waste

Ibrahim, S. (2023, August 12). *King charles lists unusual new job at buckingham palace for $63K salary.* New York Post. https://nypost.com/2023/08/12/king-charles-lists-unusual-new-job-at-buckingham-palace-for-63k-salary/

Ida, T. (2021, June 18). *Climate refugees – the world's forgotten victims.* World Economic Forum. https://www.weforum.org/agenda/2021/06/climate-refugees-the-world-s-forgotten-victims

IEA. (2019). *Emissions savings – multiple benefits of energy efficiency – analysis.* IEA. https://www.iea.org/reports/multiple-benefits-of-energy-efficiency/emissions-savings

IMPOFF. (2020, November 8). *The importance of national parks.* Impoff.com. https://impoff.com/importance-of-national-parks/

Ingram, Si. (2021, November 4). *26 hopeful visions for a sustainable future.* National Geographic. https://www.nationalgeographic.co.uk/26visions

Inspire Clean Energy. (n.d.). *What is co2e? Carbon dioxide equivalent definition, meaning & impact.* Www.inspirecleanenergy.com. https://www.inspirecleanenergy.com/blog/clean-energy-101/what-is-co2e

Ireland, P. (2022, October 13). *Ocean acidification: What you need to know.* Www.nrdc.org. https://www.nrdc.org/stories/ocean-acidification-what-you-need-know

Jackson, R. (2019, July 9). *The effects of climate change.* Climate Change: Vital Signs of the Planet. https://climate.nasa.gov/effects

Jasper, D. (2020, February 16). *Grassroots activism and how it REALLY works.* Street Civics. https://streetcivics.com/grassroots-activism-and-how-it-really-works

Jiang, X.-Q., Mei, X.-D., & Feng, D. (2016). *Air pollution and chronic airway diseases: What should people know and do?* Journal of Thoracic Disease, 8(1), E31-40. https://doi.org/10.3978/j.issn.2072-1439.2015.11.50

Keane, I., & O'Neill, N. (2023, June 7). *NYC to be clouded by Canadian wildfire smoke through Sunday.* New York Post. https://nypost.com/2023/06/07/nyc-to-be-clouded-by-canadian-wildfire-smoke-through-sunday/

Kellogg, K., & Rising, H. (2023, June 4). *How to spread awareness about climate change.* WikiHow. https://www.wikihow.com/Spread-Awareness-About-Climate-Change

Khalid, S. (2023, June 1). *10 states that banned plastic bags.* Yahoo Finance. https://finance.yahoo.com/news/10-states-banned-plastic-bags-114109220.html

Kim, D. K. (2020, January 13). *How collaboration is driving the global climate agenda.* World Economic Forum. https://www.weforum.org/agenda/2020/01/how-cross-sector-collaboration-is-driving-the-global-climate-agenda/

Konrad , K. A., & Thum, M. (2014, July 23). *What is the role of governments in climate change adaptation?* OUPblog. https://blog.oup.com/2014/07/government-adaptation-climate-change

Lindsey, R., & Dahlman, L. (2023, January 18). *Climate change: Global temperature.* Climate.gov. https://www.climate.gov/news-features/understanding-climate/climate-change-global-temperature

MacMillan, A. (2021, April 7). *Global Warming 101.* NRDC. https://www.nrdc.org/stories/global-warming-101

Make Change Staff. (2021, May 23). *Sustainability in banks: How that will change the banking sector |* aspiration. Make Change - Where Money and Mission Meet. https://makechange.aspiration.com/sustainability-in-banks

Mark, J. (2019, November 26). *Yes, actually, individual responsibility is essential to solving the climate crisis.* Sierra Club. https://www.sierraclub.org/sierra/yes-actually-individual-responsibility-essential-solving-climate-crisis

Marsh, J. (2022, September 6). *Corporate social responsibility can help companies go green.* Sustainability Times. https://www.sustainability-times.com/sustainable-business/corporate-social-responsibility-can-help-companies-go-green

Marsh, J. (2023, June 26). *6 surprising ways you can raise awareness about environmental issues.* Environment Co. https://environment.co/6-ways-to-raise-awareness-about-environmental-issues

Martin, C., & Dent, M. (2019, September 20). *How nestle, google and other businesses profit by going green.* Los Angeles Times; Los Angeles Times. https://www.latimes.com/business/story/2019-09-20/how-businesses-profit-from-environmentalism

Martinez, R., Goldblum, C., Hales, D., Monaco, M., Sthankiya, U., & Rutang Thanawalla. (2021, December 10). *Embedding climate risk into banks' credit risk management.* Deloitte Insights; Deloitte. https://www2.deloitte.com/xe/en/

insights/industry/financial-services/climate-change-credit-risk-manage
ment.html

Martins, A. (2023, February 22). *Most consumers want sustainable products and packaging.* Business News Daily. https://www.businessnewsdaily.com/15087-consumers-want-sustainable-products.html

Mary Robinson Foundation. (n.d.). *Principles of climate justice.* https://www.mrfcj.org/pdf/Principles-of-Climate-Justice.pdf

Maryville University. (2019, July 18). *Going green: Benefits of sustainability in business.* Maryville Online. https://online.maryville.edu/blog/importance-of-environmental-awareness-when-running-a-business

Maxwell, A. (2022, December 9). *Global warming impacts migration patterns.* Now. Powered by Northrop Grumman. https://now.northropgrumman.com/global-warming-impacts-migration-patterns

McAllister, S. (2022, September 27). *There could be 1.2 billion climate refugees by 2050.* Here's what you need to know. Zurich.com. https://www.zurich.com/en/media/magazine/2022/there-could-be-1-2-billion-climate-refugees-by-2050-here-s-what-you-need-to-know

Miller, M. (2019, December 14). *Exposure to nature, more empathy and less violence.* Six Seconds. https://www.6seconds.org/2019/12/14/increasing-exposure-to-nature-linked-to-empathic-behavior-less-violence

Mishra, S. (n.d.). *Jadav payeng-the forest man of india.* Retrieved August 9, 2023, from https://storymirror.com/read/english/story/jadav-payeng-the-forest-man-of-india/tuxvl6tq

Morgan Stanley Wealth Management. (2022, December 7). *How to tackle climate change in your portfolio.* Morgan Stanley. https://www.morganstanley.com/articles/how-to-combat-climate-change-in-investment-portfolio

Mulvaney, K. (2022, June 24). *What is a carbon footprint—and how to measure yours.* Environment. https://www.nationalgeographic.com/environment/article/what-is-a-carbon-footprint-how-to-measure-yours

NASA. (n.d.). *Overview: Weather, Global Warming and Climate Change.* Climate Change: Vital Signs of the Planet. https://climate.nasa.gov/global-warming-vs-climate-change

NASA. (2020, January 29). *World of change: Global temperatures.* Nasa; NASA Earth Observatory. https://earthobservatory.nasa.gov/world-of-change/global-temperatures

National Geographic. (2019, January 14). *Global warming effects.* National Geographic. https://www.nationalgeographic.com/environment/article/

global-warming-effects

National Geographic Education. (n.d.-a). *Global biodiversity.* Education.national-geographic.org. https://education.nationalgeographic.org/resource/global-biodiversity

National Geographic Education. (n.d.-b). *Niche.* Education.nationalgeographic.org. https://education.nationalgeographic.org/resource/niche

National Geographic Education. (2022a, May 20). *Greenhouse effect.* Education.-nationalgeographic.org. https://education.nationalgeographic.org/resource/greenhouse-effect

National Geographic Education. (2022b, May 20). *The importance of marine protected areas (mpas).* Education.nationalgeographic.org. https://education.nationalgeographic.org/resource/importance-marine-protected-areas

National Grid. (2023). *What are scope 1, 2 and 3 carbon emissions?* https://www.nationalgrid.com/stories/energy-explained/what-are-scope-1-2-3-carbon-emissions

Nelson, M. (2023, March 13). *How consumers are shaping corporate sustainability strategies.* Net Zero Professional. https://netzeroprofessional.com/how-consumers-are-shaping-corporate-sustainability-strategies

Nelson, R. (2019, April). *The Environmental Impact of Forest Fires.* Untamed-science.com. https://untamedscience.com/blog/the-environmental-impact-of-forest-fires

Nguyen, L. (2021, November 5). *What animals will be extinct by 2100?* Earth.org. https://earth.org/what-animals-will-be-extinct-by-2100

Nobel Foundation. (n.d.). *The nobel peace prize 2004.* NobelPrize.org. Retrieved August 9, 2023, from https://www.nobelprize.org/prizes/peace/2004/maathai/facts

Ocean and Climate Platform. (2015, March 16). *The decline of marine biodiversity.* Ocean & Climate Platform. https://ocean-climate.org/en/awareness/the-decline-of-marine-biodiversity

Ofei, M. (2019, August 25). *100+ simple tips to live a more sustainable lifestyle.* Minimalist Vegan. https://theminimalistvegan.com/live-a-more-sustainable-lifestyle

Office of Energy Efficiency & Renewable Energy U.S. Department of Energy. (n.d.). *Everyone creates a ripple: What's your ripple effect?* Energy.gov. Retrieved August 9, 2023, from https://www.energy.gov/eere/water/everyone-creates-ripple-whats-your-ripple-effect

100 companies are responsible for 71% of GHG emissions. (n.d.). Www.activesustainability.com. Retrieved August 9, 2023, from https://www.activesustainabil ity.com/climate-change/100-companies-responsible-71-ghg-emissions

100 smart ways to live sustainably. (n.d.). Sitra. https://www.sitra.fi/en/projects/ 100-smart-ways-to-live-sustainably/

100 things you can do. (2019, December 14). Sensible Sustainability. https://sensi blesustainability.home.blog/2019/12/14/100-things-you-can-do

OneTrust. (2022, August 25). *The corporate carbon footprint: A quick guide.* OneTrust. https://www.onetrust.com/blog/corporate-carbon-footprint-guide/

orbiteers. (2016, January 21). *How food choices affect the environment.* Gordon Food Service. https://gfs.com/en-us/ideas/how-food-choices-affect-envi ronment

Otte, J. (2023, July 24). *"A near-death experience": UK tourists describe escape from rhodes wildfires.* The Guardian. https://www.theguardian.com/world/2023/ jul/24/uk-tourists-rhodes-wildfires-luggage-hotels

Parker, R. R. (2022, June 14). *How climate change and environmental justice are inextricably linked.* Washington Post. https://www.washingtonpost.com/ magazine/2022/06/14/climate-justice-green-new-deal/

Patagonia. (2022). *What is worn wear?* Patagonia.com. https://help.patagonia. com/s/article/What-is-Worn-Wear

PBS. (2023, July 22). *Summer of record-breaking heat paints story of a warming world, scientists say.* PBS NewsHour. https://www.pbs.org/newshour/ science/summer-of-record-breaking-heat-paints-story-of-a-warming-world-scientists-say

Pei, A. (2021, October 7). *5 environmental benefits of sustainable transportation.* UCLA Transportation. https://transportation.ucla.edu/blog/5-environmen tal-benefits-sustainable-transportation

Pendrey, C. G., Quilty, S., Gruen, R. L., Weeramanthri, T., & Lucas, R. M. (2021). *Is climate change exacerbating health-care workforce shortages for underserved populations?* The Lancet Planetary Health, 5(4), e183–e184. https://doi.org/ 10.1016/S2542-5196(21)00028-0

Petty, L. (2022, March 11). *What are green jobs & why are they important?* The Hub | High Speed Training. https://www.highspeedtraining.co.uk/hub/what-are-green-jobs

Piraeus Bank. (2013). *Climate change and business opportunities.* https://climate-adapt.eea.europa.eu/en/metadata/case-studies/financial-institutions-preparing-the-market-for-adapting-to-climate-change-2013-climabiz/ 11261472.pdf

Polechová, J., & Storch, D. (2019). *Ecological niche - an overview.* Www.sciencedirect.com. https://www.sciencedirect.com/topics/earth-and-planetary-sciences/ecological-niche

Pruzan-Jorgensen, P. M., & Enright, S. (2018, September 20). *How businesses are collaborating for the sustainable development goals.* Www.bsr.org. https://www.bsr.org/en/blog/how-businesses-are-collaborating-for-the-sustainable-development-goals

Rao, G., & Krol, A. (2022, September 21). *Investing and climate change.* MIT Climate Portal. https://climate.mit.edu/explainers/investing-and-climate-change

Reduce Energy Use D.C. (n.d.). *Benefits of reducing green house gasses.* Reduce Energy Use DC. https://www.reduceenergyusedc.com/benefits_of_reducing_green_house_gasses?locale=en

Ricee, S. (2020, August 5). *Grassroots (what, why, how) - the complete guide to grassroots.* Diversity.social. https://diversity.social/grassroots

RTI International. (2021, April 13). *Climate justice and equity.* RTI.org. https://www.rti.org/focus-area/climate-justice-equity

Sagan, C. (1994). *Pale blue dot quotes by carl sagan.* Goodreads.com. https://www.goodreads.com/work/quotes/1816628-pale-blue-dot-a-vision-of-the-human-future-in-space

Schmidt, R. (2021, February 2). *4 reasons to live in alignment with your core values.* Life Goals Mag. https://lifegoalsmag.com/4-reasons-to-live-in-alignment-with-your-core-values

Schneider Electric. (2023, March 8). *Corporate sustainability initiatives: 5 inspiring examples.* Schneider Electric Blog. https://blog.se.com/sustainability/2023/03/08/corporate-sustainability-initiatives-5-inspiring-examples/

Shin, Y.-J. (2019). *Marine biodiversity and ecosystem services key findings from the global assessment of the intergovernmental platform for biodiversity and ecosystem services.* https://www.congress.gov/116/meeting/house/109519/witnesses/HHRG-116-II13-Wstate-ShinY-20190522.pdf

Siegler, K. (2021, August 22). *Colorado river, lifeline of the west, sees historic water shortage declaration.* NPR.org. https://www.npr.org/2021/08/22/1030154245/colorado-river-lifeline-of-the-west-sees-historic-water-shortage-declaration

Singla, A. (2020, April 24). *Does CSR help improve Brand Equity?* Www.linkedin.-

com. https://www.linkedin.com/pulse/does-csr-help-improve-brand-equity-arshiya-singla

Smith, R. (2022, October 5). *Sustainable carbohydrates: Which carbs are the least sustainable?* Vegan Food & Living. https://www.veganfoodandliving.com/why-vegan/environment/sustainable-carbohydrates/

Smith, S. (2015, April 24). *Biodiversity helps support healthy, interconnected ecosystems.* Chesapeake Bay Program. https://www.chesapeakebay.net/news/blog/biodiversity-helps-support-healthy-interconnected-ecosystems

Smithsonian Institute. (2021). *What are fossil fuels?* Ocean.si.edu. https://ocean.si.edu/conservation/gulf-oil-spill/what-are-fossil-fuels

Smoot, G. (2022). *4 main reasons why reducing your carbon footprint is important.* Impactful Ninja. https://impactful.ninja/why-reducing-your-carbon-footprint-is-important/

Society For Ecological Restoration. (n.d.). *Restoration resource center what is ecological restoration?* Www.ser-Rrc.org. https://www.ser-rrc.org/what-is-ecological-restoration

Suriyasak, C., Oyama, Y., Ishida, T., Mashiguchi, K., Yamaguchi, S., Hamaoka, N., Iwaya-Inoue, M., & Ishibashi, Y. (2020). *Mechanism of delayed seed germination caused by high temperature during grain filling in rice (oryza sativa L.).* Scientific Reports, 10(1). https://doi.org/10.1038/s41598-020-74281-9

Stallard, E. (2022, July 29). *Fast fashion: How clothes are linked to climate change.* BBC News. https://www.bbc.co.uk/news/science-environment-60382624

Sustainability For All. (2017). *100 companies are responsible for 71% of GHG emissions.* Www.activesustainability.com. https://www.activesustainability.com/climate-change/100-companies-responsible-71-ghg-emissions

sustainablesecurefoodblog. (2020, February 22). *How do my food choices affect the environment?* Sustainable, Secure Food Blog. https://sustainable-secure-food-blog.com/2020/02/22/how-do-my-food-choices-affect-the-environment

Taylor, M. (2021, August 9). *Extreme weather: How it is connected to climate change?* BBC News. https://www.bbc.com/news/science-environment-58073295

Terrascope Team. (2023, March 2). *CO2e meaning, definition, calculation, and examples.* Www.terrascope.com. https://www.terrascope.com/blog/c02e-definition-calculation-and-examples

The Dalai Lama. (n.d.). *Dalai Lama Quote. A-Z Quotes.* Retrieved August 9, 2023, from https://www.azquotes.com/quote/568553

The Editors of Encyclopaedia Britannica. (2023). *Greta Thunberg.* In Encyclopædia Britannica. https://www.britannica.com/biography/Greta-Thunberg

The Nature Conservancy. (n.d.). *The science of sustainability*. The Nature Conservancy. Retrieved August 9, 2023, from https://www.nature.org/en-us/whatwe-do/our-insights/perspectives/the-science-of-sustainability

The Nature Conservancy. (2022, August 22). *Eight ways you can reduce waste in your life and home*. The Nature Conservancy. https://www.nature.org/en-us/about-us/where-we-work/united-states/delaware/stories-in-delaware/delaware-eight-ways-to-reduce-waste

The Rewilding Institute. (2008, November 1). *What is rewilding?* Rewilding.org. https://rewilding.org/what-is-rewilding

The ripple effect – you may never know who you may impact. (2023). Boomingencore.com. https://boomingencore.com/en/article/ripple-effect-you-maynever-know-who-you-may-impact

The sustainable future of humanity: How does it look like? (n.d.). Www.greenhive.io. https://www.greenhive.io/blog/the-sustainable-future-of-humanity-whatdoes-it-look-like

The Third Pole. (n.d.). Climate refugees: *Who is a climate refugee?* The Third Pole. Retrieved August 9, 2023, from https://www.thethirdpole.net/en/hub/climate-refugees

The White House. (2022, September 1). *The rising costs of extreme weather events*. Whitehouse.gov. https://www.whitehouse.gov/cea/written-materials/2022/09/01/the-rising-costs-of-extreme-weather-events

Thorpe, A. (2022, August 11). *How to choose energy-efficient appliances – a 9-step guide*. Homesandgardens.com. https://www.homesandgardens.com/kitchens/how-to-choose-energy-efficient-appliances

Tikkanen, A. (2016, October 11). *Nature reserve* (The Editors of Encyclopaedia Britanica, Ed.). Encyclopedia Britannica. https://www.britannica.com/science/nature-reserve

Tree Canada. (2023). *National greening program*. Treecanada.ca. https://treecanada.ca/our-programs/national-greening-program/

Trendafilova, P. (2022, April 20). *Who are the companies responsible for most greenhouse gas emissions?* Carbon Herald. https://carbonherald.com/new-studyreveals-just-100-companies-have-emitted-71-of-global-emissions

Tsitsiragos, D. (2016). *Climate change is a threat – and an opportunity – for the private sector*. World Bank. https://www.worldbank.org/en/news/opinion/

2016/01/13/climate-change-is-a-threat---and-an-opportunity---for-the-private-sector

Umali-Deininger, D. (2022, March 15). *Greening the rice we eat.* Blogs.worldbank.org. https://blogs.worldbank.org/eastasiapacific/greening-rice-we-eat

UN Environment Program. (2021, August 23). *How tweaking your diet can help save the planet.* UNEP. https://www.unep.org/news-and-stories/story/how-tweaking-your-diet-can-help-save-planet

UNHCR. (2019). *Internally displaced people.* Unhcr.org. https://www.unhcr.org/about-unhcr/who-we-protect/internally-displaced-people

Union of Concerned Scientists. (2016, August 30). *The hidden costs of fossil fuels.* Union of Concerned Scientists. https://www.ucsusa.org/resources/hidden-costs-fossil-fuels

United Nations. (2022). *Biodiversity - our strongest natural defense against climate change.* United Nations. https://www.un.org/en/climatechange/science/climate-issues/biodiversity

United Nations Decade on Ecosystem Restoration. (2021). *What is ecosystem restoration?* UN Decade on Restoration. https://www.decadeonrestoration.org/what-ecosystem-restoration

United Nations Environment Program. (2019). *Climate change information sheet 22: How human activities produce greenhouse gasses.* Unfccc.int. https://unfccc.int/cop3/fccc/climate/fact22.htm

University of Colorado Boulder Environmental Center. (2020, April 22). *Why should we reduce waste?* Environmental Center. https://www.colorado.edu/ecenter/why-should-we-reduce-waste

U.S. Department of Energy Office of Legacy Management. (n.d.). *What is environmental justice?* Energy.gov. https://www.energy.gov/lm/what-environmental-justice

U.S. EPA. (2017, February 14). *US indoor water use.* Epa.gov. https://19january2017snapshot.epa.gov/www3/watersense/pubs/indoor.html

Vishnubhotla, V. (2022, February 16). *Top energy saving ways for your home.* Www.greenmatch.co.uk. https://www.greenmatch.co.uk/blog/2020/03/how-to-save-energy-at-home

Volusia County Florida. (2016, February 25). *25 ways to save water.* Volusia.org. https://www.volusia.org/services/growth-and-resource-management/environmental-management/natural-resources/water-conservation/25-ways-to-save-water.stml

Vrachovska, M. (n.d.). *8 most eco-friendly ways of transportation (short & long distance).* Almost Zero Waste. Retrieved August 9, 2023, from https://www. almostzerowaste.com/eco-friendly-transportation

Walmart. (n.d.). *Climate change.* Walmart Corporate. https://corporate.walmart. com/purpose/sustainability/planet/climate-change

Walmart. (2022). *Sustainability.* Walmart Corporate. https://corporate.walmart. com/purpose/sustainability

Waste Management. (n.d.-a). *Mixed curbside residential recycling myths.* https:// www.wm.com/content/dam/wm/documents/RecyclingResources/Posters-Guides-Tools/Recycling-Myths.pdf

Waste Management. (n.d.-b). *What is recycling & what to recycle.* Www.wm.com. https://www.wm.com/us/en/recycle-right/recycling-101

Watson, N. (2018). *Julia Butterfly Hill.* In The Editors of Encyclopaedia Britannica (Ed.), Encyclopædia Britannica. https://www.britannica.com/biogra phy/Julia-Butterfly-Hill

Western Fire Chiefs Association. (2022, July 26). *What effects do wildfires have on humans and animals?* WFCA. https://wfca.com/articles/what-effects-do-wildfires-have-humans-animals/

What is sustainable land management? (2023, March 30). Www.green.earth. https://www.green.earth/blog/what-is-sustainable-land-management

Whitten, A. (2020, January 10). *The power of the ripple effect.* The Book Refinery. https://www.thebookrefinery.com/mindset/the-power-of-the-ripple-effect

Why it's so important to reduce waste - solo resource recovery. (2019, November 21). Solo. https://www.solo.com.au/why-its-so-important-to-reduce-waste

Why marine protected areas are important for the future. (2001, December 3). Wild-coast. https://wildcoast.org/why-marine-protected-areas-are-important-for-the-future

World Migratory Bird Day. (2007). *Climate change and migratory birds.* Www.-worldmigratorybirdday.org. https://www.worldmigratorybirdday.org/ 2007/index44cb.html

World Wildlife Fund. (2022, March 15). *What is the sixth mass extinction and what can we do about it?* World Wildlife Fund. https://www.worldwildlife.org/ stories/what-is-the-sixth-mass-extinction-and-what-can-we-do-about-it

WWF. (n.d.). *Here are our top ten water-saving tips.* WWF. https://www.wwf.org. uk/what-can-i-do/top-ten-water-saving-tips

WWF. (2021a). *Effects of climate change.* World Wildlife Fund. https://www. worldwildlife.org/threats/effects-of-climate-change

WWF. (2021b). *Our planet is warming. here's what's at stake if we don't act now.*

World Wildlife Fund; World Wildlife Fund. https://www.worldwildlife.org/stories/our-planet-is-warming-here-s-what-s-at-stake-if-we-don-t-act-now

Xu, A. (2018, August 20). *5 ways the weather affects the economy.* SmartAsset. https://smartasset.com/mortgage/5-ways-the-weather-affects-the-economy

Yaniz, L. (2019, January 3). *5 ways our governments can confront climate change.* Interamerican Association for Environmental Defense (AIDA). https://aida-americas.org/en/blog/5-ways-our-governments-can-confront-climate-change

Yip, L. (2021, July 15). *Why marginalised groups are disproportionately affected by climate change.* Earth.org. https://earth.org/marginalised-groups-are-disproportionately-affected-by-climate-change/

You've heard of greenwashing, but what is youthwashing? (2023, April 5). The Climate Reality Project. https://www.climaterealityproject.org/blog/youve-heard-greenwashing-what-youthwashing

Youthwashing — climate words. (n.d.). Climatewords.org. Retrieved August 9, 2023, from https://climatewords.org/Youthwashing-1

Printed in Great Britain
by Amazon